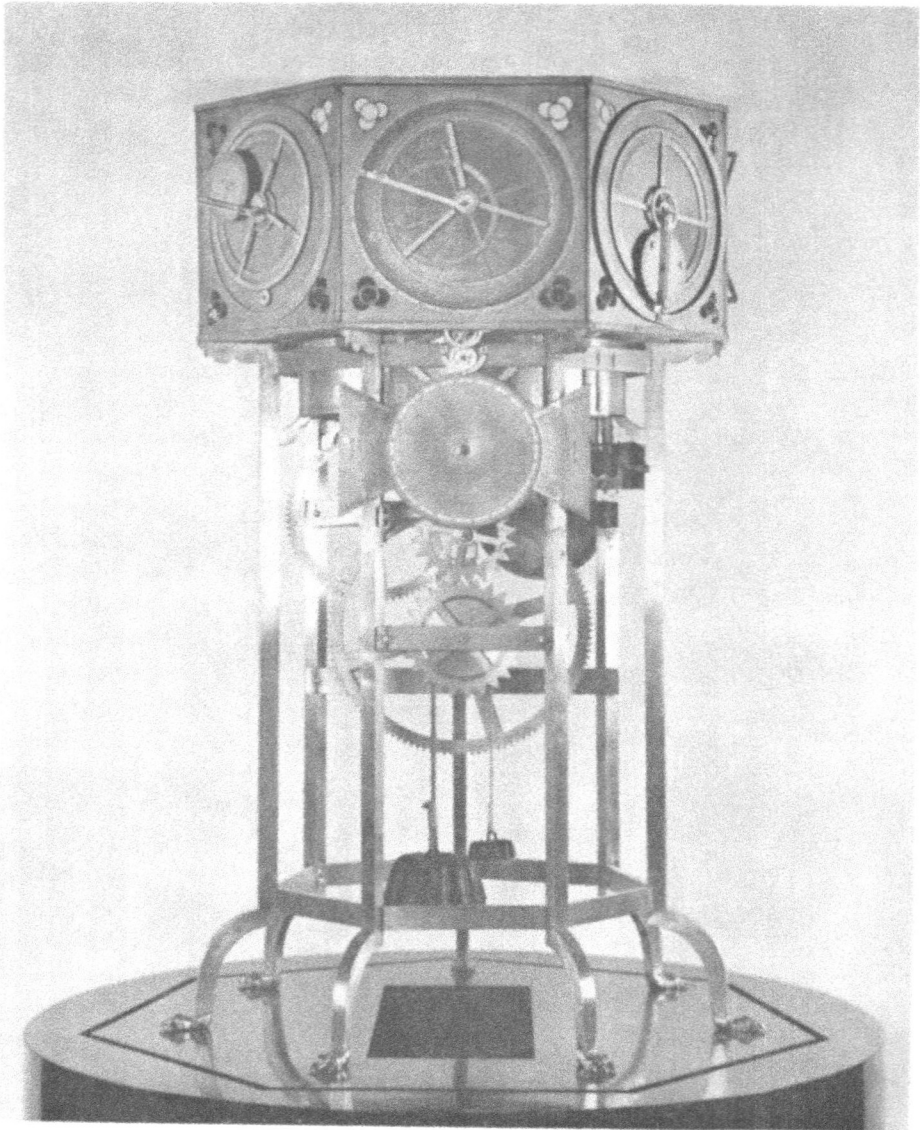

FIG. 1. Model of the astrarium of Giovanni de' Dondi, showing horary dial. In the Museum of History and Technology, Smithsonian Institution. Photograph courtesy the Smithsonian Institution.

TRANSACTIONS

OF THE

AMERICAN PHILOSOPHICAL SOCIETY

HELD AT PHILADELPHIA
FOR PROMOTING USEFUL KNOWLEDGE

———

NEW SERIES—VOLUME 56, PART 5
1966

———

MECHANICAL UNIVERSE

THE ASTRARIUM OF GIOVANNI DE' DONDI

SILVIO A. BEDINI

Assistant Director, Museum of History and Technology, Smithsonian Institution

and

FRANCIS R. MADDISON

Curator, Museum of the History of Science, Oxford University

———

THE AMERICAN PHILOSOPHICAL SOCIETY
INDEPENDENCE SQUARE
PHILADELPHIA

October 1966

Library of Congress Catalog
Card Number 66–25024

PREFACE

This publication is a conflation of studies independently begun by both authors. One of these studies, by Silvio A. Bedini, was concerned principally with the detailed history of Giovanni de' Dondi's astrarium to the end of the fifteenth century. The other, by Francis R. Maddison, on the fate of the astrarium in the sixteenth century, was originally intended to do no more than clarify an historical episode which has been often misreported. As no general survey exists of the material relating to the astrarium, and as we found ourselves able to draw attention to two hitherto neglected manuscripts and other relevant matters of interest, we have sought to make our work more generally useful by the addition of brief surveys of the extant manuscripts, of literature on de' Dondi and his astrarium, and of modern reconstructions of the astrarium. Both authors hope that the result may serve as a contribution to the sexcentenary of the completion of the astrarium, which fell in 1964.

We gratefully acknowledge assistance in research and translation, advice and suggestions which, during the preparation of this work, we have received from: Dr. J. G. G. Alexander, Department of Western Manuscripts, Bodleian Library, Oxford; Comm. Professor Dante Ballabio, Vigevano; Mr. Edwin A. Battison, Smithsonian Institution, Washington; Mr. Charles Berger, Smithsonian Institution, Washington; Professor Dott. Virginio Borroni, Direttore, Sovrintendenza ai Monumenti della Lombardia; Miss P. M. Brown, Department of Western Manuscripts, Bodleian Library, Oxford; the Honorable Earl T. Crain, U. S. Consul General, Milan; Dr. Tremblot de la Croix, Conservateur Bibliothèque de l'Institut de France, Paris; Miss Carolyn Fawcett, Radcliffe College, Cambridge; Professor Cav. Attilio Ferrar-Trecate, Vigevano; Mr. Jack Goodwin, Smithsonian Institution Library, Washington; Dr. C. H. Josten, Curator Emeritus, Museum of the History of Science, Oxford; Dr. Harriet Pratt Lattin, Belvedere-Tiburon, California; Dr. R. E. W. Maddison, Heston; Dr. C. Donald O'Malley, School of Medicine, University of California at Los Angeles; Dr. A. C. de la Mare, Department of Western Manuscripts, Bodleian Library, Oxford; Professor Dott. Alfeo R. Natale, Direttore, Archivio di Stato di Milano, Milan; Dr. Joseph Needham, F. R. S., Master of Gonville and Caius College, Cambridge; Dr. J. G. North, Nuffield Research Fellow in the History and Philosophy of Science, University of Oxford; Professor Carlo Pedretti, Department of History, University of California at Los Angeles; Dott. Emma Pizzoni, Milan; Professor Derek J. de Solla Price, Department of the History of Science, Yale University; Dr. Tadeusz Przypkowski, Panstwowe Museum im Przypkowski, Jedrzejow; Dr. Ladislao Reti, Sao Paulo; Mr. W. F. Ryan, Museum of the History of Science, Oxford; Mrs. Catherine S. Scott, Smithsonian Institution, Washington; Rev. Robert Stenger, O. P., Dominican House of Studies, Washington; Professor Lynn White, Jr., Department of History, University of California at Los Angeles; Dr. Maria Zakrzewska, Museum Universytetu Jagiellonskiego, Cracow. We are also indebted to Gerda H. Bedini for the compilation of the index.

S. A. B.
F. R. M.

Washington and Oxford
April, 1966

3

MECHANICAL UNIVERSE
The Astrarium of Giovanni de' Dondi

SILVIO A. BEDINI AND FRANCIS R. MADDISON

"Using his keen mind, he built a perfect machine, where the intricate mass of the orbs and planets is clearly and distinctly known to be moved in an orderly manner, so that it seems to be a divine rather than a human work." [1]

CONTENTS

INTRODUCTION

One of the most important machines of all time was the complicated astronomical clockwork called the "astrarium" or "planetarium" which was designed and constructed between 1348 and 1364 by Giovanni de' Dondi, professor at the University of Padua.[2] Contemporary accounts relate that de' Dondi devoted sixteen years of his leisure time to the project, which included preparatory astronomical studies and the design and construction of the mechanism, said to have been accomplished with his own hands. The completed machine was a masterpiece unsurpassed in its own time. Today, six centuries later, it remains one of the greatest achievements combining science and technology.

The importance attributed to the astrarium through the centuries has steadily increased rather than diminished with the passage of time. In spite of the numerous published works which have described or mentioned it, however, much remains to be learned about this mechanism. Many questions remain unanswered about the circumstances relating to its invention, as well as about various aspects of its production and assembly.

The construction of each part of the mechanism was described in considerable detail by Giovanni de' Dondi

in a manuscript work entitled *Tractatus astrarii*, written at about the time of the astrarium's completion. This work was recopied at various times in manuscripts, sometimes bearing the title *Opus planetarium*. There are, however, substantial variations in these texts relating to details of the mechanism and these have added to the confusion. At least eleven of these manuscript copies have survived and are now known to scholars (see below, p. 40).

The present work has been undertaken for the purpose of recording the history of the astrarium from the time of its construction to its disappearance more than one hundred fifty years later. An attempt is herein made to collect and systematize in chronological order all available surviving contemporary accounts and later records relating to the astrarium. Many of the documents here presented for the first time have come to light in regional archives; others have been previously noted only in histories of local art and architecture.

THE SIGNIFICANCE OF THE ASTRARIUM

The astrarium of Giovanni de' Dondi was probably not a unique fourteenth-century achievement. Other early fourteenth-century clocks certainly indicated astronomical data. The earliest astronomical clock of which we have any record is that which was devised *ca.* 1327–1330 by Richard of Wallingford (1292?–1336) for the Abbey of St. Albans in Hertfordshire. The

---

[1] [De' Dondi, ——], *Familia nostra* [144], (Appendix I, Document I). Numbers in brackets refer to items in bibliography.

[2] Baillie, "Giovanni de' Dondi" [81]. The most complete published description in English of the astrarium is to be found in Lloyd, "Giovanni de Dondi's Horological Masterpiece 1364" [97].

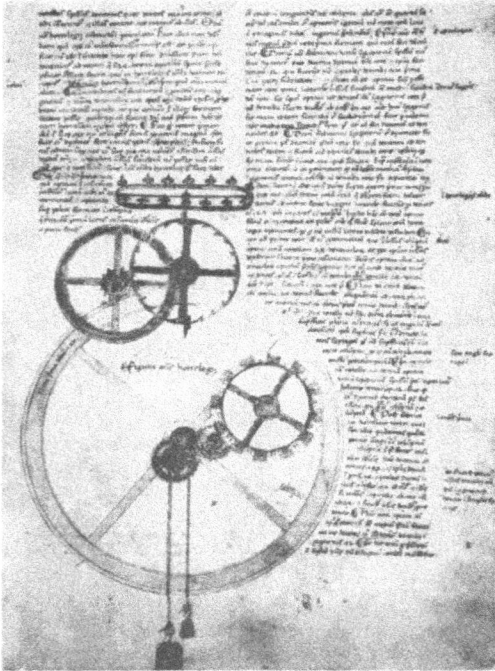

FIG. 2. Schematic drawing of the movement of the astrarium. Reproduced from fol. 3ʳ of Cod. D. 39, *Tractatus astrarii*, Biblioteca Capitolare Vescovile, Padua.

orphaned son of a smith of Wallingford, Richard was adopted by William de Kirkeby, Prior of Wallingford, and went to study at Oxford. Six years later, he entered the Monastery of St. Albans. After three years, he was sent back to Oxford by the Abbot, Hugh de Eversdon, and spent nine years in the study of philosophy and theology. In later years, he is said to have regretted devoting more time to scientific studies than to theology. The loss to theology, however, was a considerable gain to medieval astronomy and technology.

After the death of Abbot Hugh, in September 1327, Richard was elected Abbot of St. Albans. According to the *Gesta Abbatum Monasterii Sancti Albani*, written just over half a century later by Thomas Walsingham, Precentor of the monastery,

... he [Richard] produced many books and instruments relating to astronomy, geometry and certain other sciences, in which he excelled all others of his time.

Among these, was a remarkable astronomical instrument, the like of which had never before been seen, called "Albion," meaning all-by-one. . . .[3]

[3] Riley (ed.), . . . *Gesta Abbatum* . . . [62a] 2: p. 207 (original Latin text). The translation is slightly modified from that of Baillie, *Watches* [6], p. 30.

The Albion, described in a treatise[4] written by its inventor, was a complicated astronomical instrument, basically an equatorium. It was one of the most influential astronomical computing devices of the Middle Ages. Richard of Wallingford also invented the rectangulus, which was inspired by the torquetum, and introduced new trigonometrical methods.

It must have been shortly after his election as Abbot that Richard began the construction of an astronomical clock for the Abbey. Information about this clock has been extremely meager, and has had to be pieced together from a number of isolated references. A miniature painting in a manuscript, in the British Museum, listing the benefactors of the Abbey of St. Albans depicts Richard and his astronomical clock, but it is not very informative and is probably inaccurate.[5] Two passages in the *Gesta Abbatum,* however, mention the clock. Recounting the history of Richard's abbacy, Thomas Walsingham says:

He [Richard] made a noble work, a horologium, in the church, at great cost of money and work; nor did he abandon finishing it because of its disparagement by the brethren, although they, wise in their own eyes, regarded it as the height of foolishness. He had, however, the excuse that he originally intended to construct the horologium at less expense, in view of the great and generally recognised need for repair of the church, but that in his absence, and as a result of interference by some brethren and the greed of the workmen, it was begun on a costly scale and it would have been unseemly and shameful not to have finished what had been put in hand.

Indeed, when on a certain occasion, the very illustrious King Edward the Third came to the monastery in order to pray, and saw so sumptuous a work undertaken while the church was still not rebuilt since the ruin it suffered in Abbot Hugo's time, he discreetly rebuked Abbot Richard in that he neglected the fabric of the church and wasted so much money on a quite unnecessary work, namely, the above-mentioned horologium. To which reproof, [the Abbot] replied, with due respect, that enough Abbots would succeed him who would find workmen for the fabric of the monastery, but that there would be no successor, after his death, who could finish the work that had been begun. And, indeed, he spoke the truth, because in that art nothing

[4] Almost thirty complete texts or fragments of this treatise are now known. A transcript, by the Rev. H. Salter, of the second part only, based on MSS Corpus Christi College 144 and Laud Misc. 657 (both in the Bodleian Library, Oxford), was published by R. T. Gunther, *Early Science in Oxford* [38a] 2: pp. 349–370. The incipit of the Albion treatise is: *Albyon est geometricum instrumentum* . . . ; in the Laud MS, however, the initial paragraph is misplaced and the beginning is therefore misleading. See also note 14, below.

[5] MS Cotton Nero D. vii, bottom left-hand corner of f. 20ʳ. This manuscript has no title, but was described by Smith in his *Catalogus librorum manuscriptorum Bibliothecae Cottonianae* [114a], p. 57, as "Catalogue Benefactorum & omnium eorum, qui in plenum fraternitatem Monasterii S. Albani recepti erant: cum compendiariis historiis eorundem, & elegantissimis picturis. Ex dono vicecomitis S. Albani, anno 1623." F. 46ʳ indicates that the list of abbots, etc., was completed in 1484 under the abbacy of Thomas Ramryge.

of the kind remains, nor was anything similar invented in his lifetime.[6]

The other relevant passage in the *Gesta Abbatum* refers to the abbacy of Thomas de la Mare, 1349 to 1396:

The upper dial and wheel of fortune, originally arranged by Abbot Richard, that master of all such things, but meanwhile set aside by reason of his early death and other more urgent expenditure, was nobly completed, at the charge and by the exertions of this Abbot [Thomas de la Mare], and by the work of Master Laurentius de Stokes, an eminent *horologiarius,* and of one of his brother monks, called William Walsham, who in the work of their hands and in

[6] Riley (ed.), . . . *Gesta Abbatum* . . . [62a] 2: pp. 281–282 (original Latin text). The translation is based on that of Baillie, *Watches* [6], pp. 29–30, which we have changed in places after comparison with the original.

FIG. 3. Miniature painting depicting Richard of Wallingford and his astronomical clock constructed for the Abbey of St. Albans, ca. A.D. 1320–1330. The miniature occurs in MS. Cotton Nero D. vii, in the bottom left-hand corner of f. 20ᵛ. This manuscript, now in the British Museum, has no title but was described by Thomas Smith in his *Catalogus librorum manuscriptorum Bibliothecae Cottonianae* (Oxford, 1696, p. 57) as "Catalogus Benefactorum & omnium eorum, qui in plenam fraternitatem Monasterii S. Albani recepti erant: cum compendariariis historiis eorunden, & elegantissimis picturis. Ex dono Vicecomitis S. Albani, anno 1623." F. 46ᵛ indicates that the list of abbots, etc., was completed in 1484 under the abbacy of Thomas Ramryge. Photograph courtesy the Trustees of the British Museum.

skill of fashioning almost surpassed all the craftsmen of the district. The cost of this, on account of the size and complexity of the work, was estimated at a hundred marks and more.[7]

Either the King's displeasure, or Richard's increasing infirmity resulting from the leprosy with which he was afflicted, must have prevented the completion of the clock in Richard's lifetime. This does not necessarily mean that the clock did not function before he died. The upper dial and the "wheel of fortune" could have been further elaborations of the clock which Richard had begun. The "wheel of fortune" may have been an automaton, or merely a rotating pictorial disc.[8] The *Gesta Abbatum* mentions that in 1394 the canonical hours were replaced by 24-[equal] hour reckoning.[9] Perhaps this was made possible by the completion of Richard's clock, but any astronomical clock would presumably have indicated equal hours, whatever else it showed.

A later description of Richard of Wallingford's clock was given by the antiquary John Leland (1506?–1552), who may have seen the clock about 1540:

. . . And indeed, when it became possible through sufficient means, he wished to show a miracle by some great work not only of inventiveness, but also of learning and excellent craftsmanship. Therefore, he constructed such a fabric of a horologium, with great labour, at greater cost, and with still greater skill and, in my opinion, it has not its equal in all Europe; one may look at the course of the sun and moon or the fixed stars, or again one may regard the rise and fall of the tide, or the lines with their almost infinite variety of figures and indications. And when this task, worthy of history, had been brought to a conclusion, since he was easily the greatest mathematician of all in his time, he wrote the *Canones* in which book the work was set down so that such a remarkable mechanism should not become worthless through any error of the monks or remain silent because of the unknown nature of its structure.[10]

One would indeed have expected that, as Leland says, Richard of Wallingford would have written an account of his astronomical clock, as he wrote on his Albion and his rectangulus. No such account, however, was known to be extant. In 1956 Professor Derek J. de Solla Price reported his discovery of a text which describes an astronomical clock in a manuscript of about 1450, formerly in the Abbey of St. Albans.[11] The manuscript, now belonging to Gonville and Caius College, Cambridge, appears to be a collection of various texts made

[7] Riley (ed.), . . . *Gesta Abbatum* . . . [62a] 3: p. 385 (original Latin text). The translation is based on that of Baillie, *Watches* [6], p. 30, corrected against the original.

[8] The Wheel of Fortune was a common medieval symbol representing the variability of good fortune, e.g., the rise and fall of temporal power. For a drawing of a Wheel of Fortune, ca. 1225–1250, see Bowie (ed.), *The Sketchbook of Villard de Honnecourt* [18], frontispiece.

[9] On the use of canonical hours in relation to the early history of clocks, see Edwardes [33a], pp. 2 ff.

[10] Original Latin text in Tanner [69], p. 629. The translation based on that of Baillie, *Watches* [6], pp. 30–31, amplified and modified after consulting the original.

[11] Price, "Two Medieval Texts" [107], p. 156.

by a former Abbott of the Abbey, including copies from manuscripts which were available there.[12] Price noted that the volume contained a text entitled: *Inc. conclusiones diuidendi rotas pro horologia astronomico pro motibus planetarum* . . . ("Here begins a text on the division of the wheels for an astronomical clock showing the motions of the planets . . ."). The text begins with the words, *Habita rota certi numeri* . . . (A given wheel of a certain number of teeth . . ."), and consists of a total of about 1,300 words.[13] It was Price's opinion that this might be a copy of a fragment of Richard's missing work describing the clock he built at St. Albans.

Recently, Dr. J. G. North, of Oxford, has drawn attention [14] to a manuscript in the Bodleian Library, Oxford,[15] which contains an appreciably longer, and probably nearly complete version of the text discussed by Price. The Oxford text, however, presents certain problems which have not yet been fully resolved.[16] The text includes four or five illustrations. Three of these illustrations show trains of gears or other mechanical linkages; one of them is a sectional view through the "dial," the indexes, and the rotating lunar globe of an elaborate astronomical clock. The text describes the representation by wheel-work of the motions of various heavenly bodies, including the calculation—with the aid of tables provided—of the number of teeth required on the gear-wheels. It also tells how to make the clock ring a bell. The passage dealing with the manner in which the clock was controlled is exceedingly vague. Certainly, weights were used in the going as well as the striking train. So far, we have not been able to trace any other text of this treatise.[17]

The Oxford text is found in a manuscript which contains other medieval astronomical texts, including some by Richard of Wallingford, notably his treatise on the Albion. This manuscript, like that discussed by Price, was associated with the Abbey of St. Albans, but is probably about a century earlier. A preliminary study suggests a date before 1360. It is reasonable to suppose that this text describes Richard of Wallingford's clock, and that it is probably Richard's own treatise on the subject.[18]

That Richard of Wallingford's astronomical clock bore some mechanical resemblances to the astrarium of Giovanni de' Dondi is likely. Two medieval designers, given a similar problem and similar technological re-

sources, are likely to have produced somewhat similar devices. In details, however, especially in the arrangement of the dials, the two clocks were probably very different. One might assume that, in order to produce an astronomical clock with any pretention at accuracy, Richard of Wallingford must, like de' Dondi, have known and used a mechanical escapement with his weight drive, thus narrowing the gap of time between the probable invention of the escapement and its earliest known applications in Europe.[19]

Although the astronomical clock at the Abbey of St. Albans and certain other fourteenth-century clocks of which we have record are earlier than de' Dondi's astrarium, it remains true that the latter is the earliest clockwork of which an almost complete description and incontestable documentation have survived. It is, furthermore, one of the most complicated clockworks known to have been produced until recent times. The descriptions given by Richard of Wallingford and by de' Dondi of their astronomical timepieces are the more valuable in that no examples of clockwork are known to have survived from the early or middle fourteenth century. The earliest extant mechanical clock is probably that of Salisbury Cathedral, *ca.* 1390. This clock is of the type of turret clock common thereafter and, though of the greatest interest as regards its mechanical construction, tells us nothing about the nature or appearance of the first astronomical clocks. Perhaps the only

---

[12] MS 230/116.

[13] Ff. 11^v–14^v (pp. 31–36).

[14] Private communications, May 1965–April 1966. Dr. North is engaged on a comprehensive edition of the complete works of Richard of Wallingford which will contain translations and commentaries.

[15] MS Ashmole 1796.

[16] The manuscript does not give the text entirely continuously, or always in the correct sequence.

[17] The correct incipit appears to be: *Habita rota certi numeri*. . . .

[18] An advance notice of this text is in preparation by Dr. North and one of the present writers (F. R. M.).

[19] See below, p. 62.

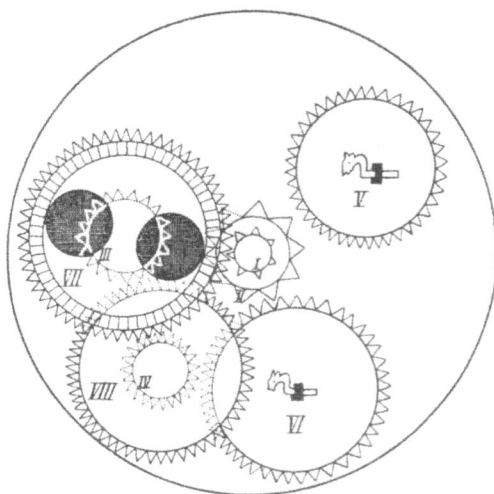

Fig. 4. Drawing of calendrical gearing described by al-Bîrûnî about A.D. 1000. Reproduced from E. Wiedemann, "Ein Instrument" [117b], p. 8.

FIG. 5. Persian astrolabe, of brass, with geared calendar movement, made by Muḥammad b. Abî Bakr of Iṣfahân in A.D. 1221–2. Front, showing *rete*. The two large lugs on the *rete* serve as sights, the calendar on the back of the instrument preventing the use of a normal alidade. The calendar was probably operated by turning the horse (wedge) and pin. Museum of the History of Science, Oxford, No. IC 5. Photograph Edmark, Oxford.

surviving fragment of a fourteenth-century astronomical clock is the *rete* of a late Gothic astrolabe, ?*ca.* 1400, now in the Museum of the History of Science, Oxford.[20]

Richard of Wallingford's clock may have been a mechanical equivalent of his Albion, and de'Dondi's astrarium provides tangible evidence in confirmation of the theory presented by Price[21] that the time-telling mechanical clock originally had no affinity with common time-measurement devices such as the sundial, sand glasses, and clepsydrae, but that the mechanical clock was in fact derived from the mechanized astronomical devices which he calls "protoclocks." The completion of the astrarium in the fourteenth century may be linked continuously backward through history to the first appearance of simple gearing. Price suggests that the origin of mechanized astronomical devices, such as the astrarium, may well have been the Islamic tradi-

tion of equatoria which gradually migrated to the Western world, in particular through the descriptions of instruments in the *Libros del saber* of Alfonso the Wise (*ca.* 1276). He points out that concrete evidence of this migration is to be found in the fact that the dials and complicated gearwork of de' Dondi's astrarium constitute a series of equatoria. De' Dondi stated that this was in fact what he had done, basing the series on the instructions provided by Campanus of Novara in

FIG. 6. Persian astrolabe, of brass, with geared calendar movement, made by Muḥammad b. Abi Bakr in A.D. 1221–2. Back, showing calendar. The circular opening reveals a lunar phase diagram, which may be compared with the lunar volvelles found on many old European horological instruments, and even today on the dials of long-case clocks. The *abjad* numeral visible in the small rectangular opening gives the age of the moon (and, therefore, the date in a lunar calendar). Below, within a zodiacal calendar scale, are two concentric rings, the outer ring inset with a small gold disk representing the sun, the inner ring formerly having a similar inset (probably of silver) representing the moon. The rotation of these rings show the relative positions of the sun and moon (opposition, conjunction, etc.) and the position of the sun in the zodiac. The inscription across the middle of the back is the maker's signature, and reads, "Made by Muḥammad b. Abi Bakr b. Muḥammad ar-Râshidi al-Ibari [or al-Abiri] al-Iṣfahâni." The instrument is dated "618 [A.H. = A.D. 1221–2.]," at the end of the long inscription around the edge. Museum of the History of Science, Oxford, No. IC 5. Photograph Edmark, Oxford.

[20] See fig. 38 and note 131 below. *Cf.* also the remarks on the dial of Jacobo de' Dondi's clock, p. 17 below.

[21] Price, *Origin of Clockwork* [109], pp. 103–105.

his *Theorica planetarum* (see below, pp. 15–16). These are mechanized in the same manner as the well-known calendrical device described by al-Birûni in about A.D. 1000 which was subsequently employed in the geared astrolabe of Muḥammed b. Abi Bakr of Iṣfahân of A.D. 1221–1222, in the Museum of the History of Science at Oxford.

A related instrument, somewhat later in date, is a geared astrolabe of early Gothic design, in the collection of the Science Museum, London. French in origin, this astrolabe dates from *ca.* 1300. The movements and cor-

Fɪɢ. 7. Persian astrolabe, of brass, with geared calendar movement, made by Muḥammad b. Abi Bakr, A.D. 1221-2. Inside of back plate, showing gear train, similar to that described by al-Birûni, *ca.* A.D. 1000 (*cf.* fig. 4). The gear train count is as follows: 48 − 13 + 8 − 64 + 64 + 10 − 60. The pinion of 8 teeth has been incorrectly replaced by a pinion of 10. The gear of 48 teeth should make 13 (lunar) rotations, while the double gear of 64 + 64 makes 6 revolutions of double months (of 29/30 days) and the gear of 60 makes a single turn in the hegiral year of 354 days. Museum of the History of Science, Oxford, No. IC 5. Photograph Edmark, Oxford.

rect relative positions of the sun and moon are indicated by pointers moved by an arrangement of gears.

The complicated achievement of the astrarium seems to verify Price's assumption that the ordinary mechanical clock was a degeneration from the models of the universe in miniature based on devices such as the astrolabe and the equatorium.

These astronomical models, if mechanical, probably could not have been powered by clockwork before *ca.* 1270.[22] Nevertheless, it is known from the existence of

[22] Thorndike, "Invention" [117a]; Thorndike, *Sphere* [70a], pp. 179–180 (Latin text), 229–231 (English translation, not in Thorndike [117a]); Price, *Origin of Clockwork* [109], p. 106.

Fɪɢ. 8. French gothic astrolabe, of brass, *ca.* 1300. Front, showing *rete*. The relative motions and positions of the sun and moon are indicated by pointers, moved by a geared arrangement. The pointer of the moon indicator is missing. The Science Museum, London. No. IC 198. Photograph courtesy the Science Museum.

several fairly elaborate monastic water-clocks of earlier date that their mechanisms could have been quite complex. Consequently, de' Dondi's astrarium appears as one of the earliest devices combining the tradition of the astronomical models of the universe with the fairly recently invented clockwork. It is a matter of considerable significance that, in his treatise on the astra-

Fɪɢ. 9. French gothic astrolabe of brass *ca.* 1300. Train of cogwheels of pointer for the sun. The Science Museum, London. No. IC 198. Photograph courtesy the Science Museum.

rium, Giovanni de' Dondi provided no details for the installation of the horary dial. He referred to it as an ordinary or "common" clock, and he went on to add that if the reader was incapable of producing this part of the mechanism, he would be wise not to attempt to make the rest of it.

The importance of this statement must not be overlooked, indicating as it does that clockwork was so

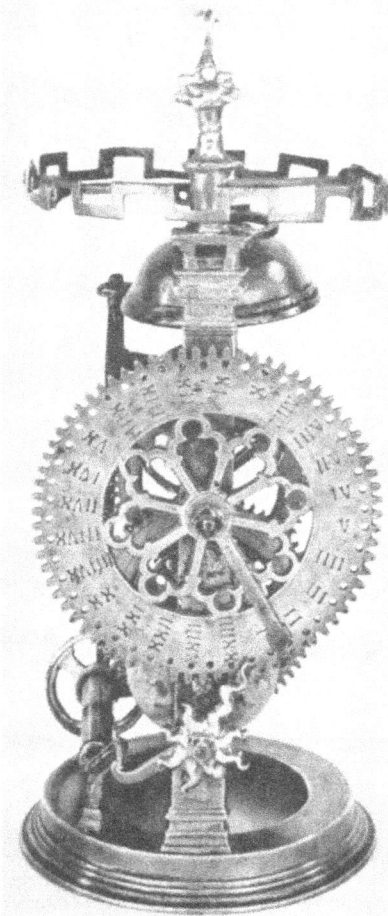

FIG. 10. Alarm clock, of brass, ?ca. fifteenth century. Said to be the earliest known surviving domestic clock. The base, mountings and wheels are of brass, and the arbors and ratchets are of iron. Overall height 25 cm. The great wheel makes one revolution per hour, and carries on its arbor a three-leaved pinion which drives the dial wheel, which makes one revolution in 24 hours. Owned by Captain Antonio Simoni, Bologna. See Bedini [82]. Photograph courtesy the Science Museum, London.

FIG. 11. Alarm clock depicted in an intarsia panel executed between 1528 and 1543 by Fra Damiano Frambelli da Bergamo. In the Basilica of San Domenico, Bologna. Photograph courtesy Capt. A. Simoni, Bologna.

common and familiar that de' Dondi considered that no description was required. Yet, clockwork could hardly have existed prior to 1270, less than a hundred years before the completion of de' Dondi's astrarium.

Giovanni de' Dondi's astrarium must be considered important for another reason. This masterpiece, in common with other "protoclocks" is part of that tradition which provided the inspiration for the astronomical public clocks on European churches and public buildings which became so popular during the centuries that followed.

Finally, the technical descriptions of the astrarium which have been preserved in the several manuscripts of the *Tractatus astrarii* and *Opus planetarium* which have survived, provide an invaluable record of the development of astronomical science and of the technology achieved in Europe by the mid-fourteenth century.

THE INVENTOR

Giovanni de' Dondi (1318–1389) was born in Chioggia, the son of Jacopo de' Dondi, the municipal physician. In 1349 Jacopo transferred his entire family to

Padua and established residence opposite the Pozzo Mendolo.

Following his father's profession, Giovanni became professor of medicine at the University of Padua in 1350 or 1352. In 1359 he was appointed a member of the four faculties of medicine, astrology, philosophy, and logic. He became widely recognized as a scholar of wide learning. He lectured in medicine at Florence from early 1368 through 1370. In 1371 he served as ambassador to Venice and in the following year he was a member of a committee of five citizens appoined to establish boundaries between Carrara and the Venetian Republic. In the same year he was among those citizens of Padua who voted to wage war against the Venetians.

Giovanni de' Dondi was married twice. In January, 1354, he became free of paternal jurisdiction, and in September of the same year he married Giovanna di Reprandino dalle Calze. He was the father of a son, Jacopo, and four daughters. He was widowed and

FIG. 12. Alarm clock depicted in intarsia panel of the early sixteenth century. Believed to have been originally installed in a church in Bologna, it is now in the collection of the Victoria and Albert Museum, London. Photograph by permission of the Trustees of the Victoria and Albert Museum.

FIG. 13. Gothic weight-driven domestic clock, of iron, *ca.* 1500. Museum of History and Technology, Smithsonian Institution, Washington. Photograph courtesy the Smithsonian Institution.

subsequently remarried, in July 1379. His second wife was Caterina di Gerardo da Tergola, and he became the father of four more children.[23]

After the conflict with Venice, de' Dondi appears to have lost the friendship of his former patron, Francesco da Carrara, Prince of Padua. Meanwhile, he was befriended by Gian Galeazzo Visconti of Pavia who loaded him with benefits and honors. De' Dondi was appointed to the faculty of the Università of Pavia, and in August 1382 he was residing in the Visconti palace under the loggia of the chancery. He established official residence in Pavia in September 1387. When in his last years he visited Antonio Adorno, the Doge of Genoa, he apparently became ill and died at Milan on 22 June 1389.

His body was interred in the family vault in Padua,

23 Gloria, *Monumenti* [38], pp. 381–386.

near his father, and his final resting place was marked
with an epitaph:

Tumbica sic celsum conclusit petra Joannem
　Quem Dondis satum genuisti Clugia lymphis
Ast animum medio separarunt numina Jano.
　Hic recubans Patavi. supremaque sidera terris
Caelicis hic septem confecit mystica stellis.
　Praeteriti doctor renovans hic sanguinis umbram
Philosofus. Rhetor. Medicus mortalis in orbe
　Opera terrenis liquit super aethera scandens.
M. CCC. LXXXX. die XXVII Septemb."

Following is a free translation:

This tombstone covers the eminent Johannes,
Whom you, Chioggia, have brought forth, filled with the
Dondi waters
But like Janus the gods severed his mind.
Here [he is] resting at Padua. On earth, here,

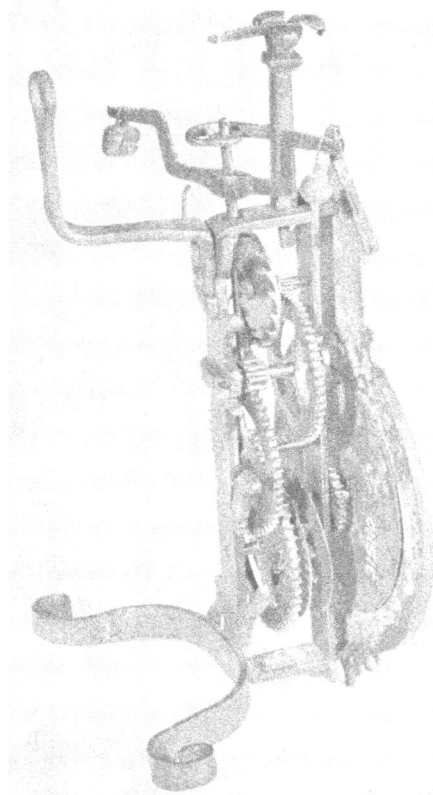

FIG. 14. Side view of Gothic domestic clock, *ca.* 1500, showing
details of movement and verge and foliot escapement. Museum of History and Technology. Photograph courtesy the
Smithsonian Institution.

FIG. 15. Statue of Giovanni de' Dondi. Erected in the Prato
delle Valle, Padua, in 1778 by his descendants Giovanni
Antonio and Cav. Francesco Dondi dall'Orologio. The
inscription on the base is "JOHAN. DE DONDIS/AB
HOROLOGIO/PHILOSOPHO ET MATEMATICO/
POSTERI EJUS/AN. MDCCLXXVIII." Photograph
courtesy the Ente Provinciale per il Turismo, Padua.

He represented the mystical uppermost stars by seven
heavenly stars.
A doctor, he renews the shadow of past blood [his
ancestors'?].
A doctor, a philosopher, and a rhetor,
A mortal physician in the world,
He leaves on earth works rising [in fame] above the ether.
27 September 1390.

After his death an inventory of his personal property
was filed by his widow.[24] In addition to other possessions, the inventory listed the volumes in his private
library. Among these were his personal copies of several versions of his manuscript on the construction of
the astrarium. From these it is apparent that the name
"astrarium" originated with Giovanni, or in his time, to
describe the masterpiece. His library contained a copy
of *Planetarium dicti Mag. Jo.* (*Planetarium of the said
master Giovanni*), and another work listed as *Pars
Planetarii.* Although these may have been copies of
his father's astronomical tables which bore the same

---

[24] Lazzarini, *Libri* [96], pp. 26–32.

title, it is possible that they were manuscripts of the *Opus planetarium* relating to Giovanni's clockwork.

Another intriguing entry in the inventory of de' Dondi's library is No. 58, "Primo quaterni quinque de opere horologii sui." These "five notebooks of his horological works" may possibly have consisted of the working notes he compiled during the sixteen years he was engaged in the design and the construction of his astrarium, and on which his *Tractatus astrarii* would have been based.

Most interesting among the manuscripts of Giovanni's personal library is one described as *Tractatus aliqui per eundem Mag. Jo. composti* and another entitled *Astrarium Jo. de Dondis.*[25] At least one of these, presumably the latter, remained in the family until it was deposited in the Biblioteca Capitolare Vescovile of Padua in 1795 by Monsignor Francesco Scipione, Marchese Dondi dall'Orologio.

DESCRIPTION OF THE ASTRARIUM

The astrarium was designed to indicate the motions of the planets according to the Ptolemaic theory, that is, of the sun, the moon, and the five planets then known, in addition to the mean and sidereal times. It was made in two sections on a heptagonal framework, the upper part of which consisted of seven large dials, one each for the sun, moon, and five planets. On the lower framework were a 24-hour dial, a dial for the fixed feasts of the church, another for movable feasts, and one for the nodes or points of intersection of the orbits of the sun and moon. Wings attached at either side of the horary dial indicated the times of the rising and setting of the sun in Padua. These tables are divided on each side into months and days in accordance with the dates of the summer and winter solstices in the Julian calendar, with December 13 through June 12 on the outside and June 13 through December 12 on the inside.

De' Dondi employed 12:00 noon instead of midnight for the point of reference on which his astronomical observations were based. Accordingly, the dividing line placed below the meridian was used to indicate the passage of the hours from the previous noon or point of reference. The clock dial revolved counterclockwise, making it necessary to read from the left edge of the horary divisions.

The clock was regulated by a balance wheel, shaped like an elaborate crown, which beat two seconds. The description furnished by de' Dondi in his manuscript indicated that this was the usual rate of the common clock, and he recommended adjustment by means of weights attached to the balance if it were fast, and if it were slow by the addition of weights to the single driving weight, which provided the motive power for the mechanism.

The annual calendar dial with fixed feasts consisted of

[25] *Ibid.*

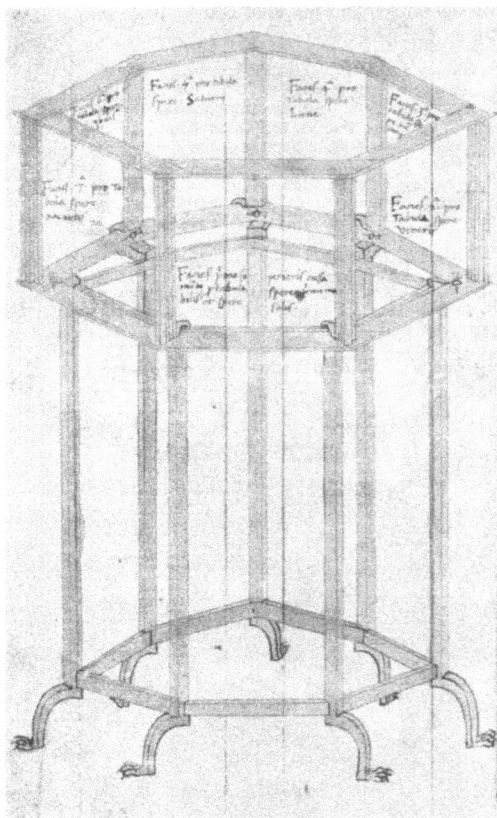

Fig. 16. Drawing, 1461, of the framework of the astrarium from MS. Laud Misc. 620, folio 7ᵛ. Bodleian Library, Oxford. Photograph courtesy the Curators of the Bodleian Library.

a large circular band revolving in the horizontal plane and filling almost the entire width of the frame. The top edge was serrated with 365 teeth corresponding to the days of the year. The outer side of the band was engraved with indications of the length of each day of the year in hours and minutes, the name of the saint to be commemorated, the date of the month and the dominical letter, the relevant column of which was visible through an opening in the dial plate.

The calendar of movable feasts consisted of three linked chains. The topmost chain had 28 links and showed the dominical letter [26] over a period of 28 years.

[26] *Dominical letter:* The first seven letters of the alphabet, A through G, are assigned in consecutive order to the first seven days of the month of January. The letter which coincides with the first Sunday within this period in a given year is called the dominical letter, and is used for ascertaining the date of Easter Sunday. The date of Easter regulates the dates of other

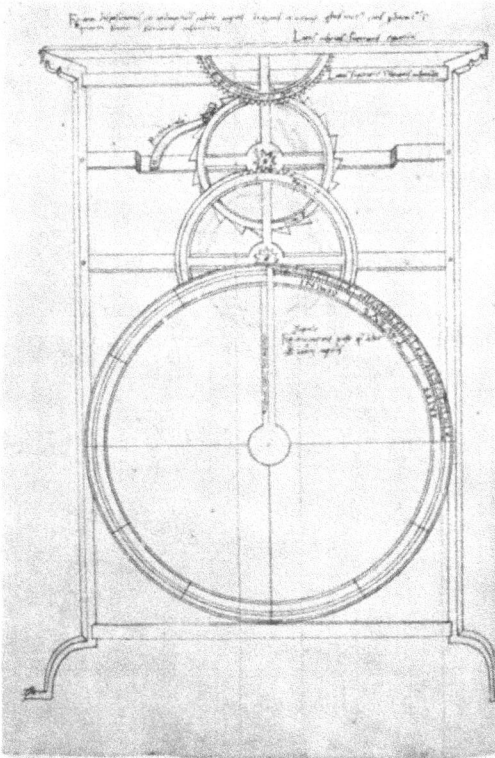

FIG. 17. Drawing, 1461, of the dial of the nodes on the astrarium from MS. Laud Misc. 620, folio 87ᵛ in the Bodleian Library, Oxford. Photograph courtesy the Curators of the Bodleian Library.

The middle chain was wider and had 19 links for the lunar cycle indicating the movable feasts. The lowermost chain had 15 links and indicated the period of indiction.[27]

The five planets of which the motions were indicated on the upper structure were Mars, Mercury, Jupiter, Venus, and Saturn, driven from teeth on the wheel of the annual calendar ring. The dials of Venus, Jupiter, and Mars were driven from the calendar wheel of the annual calendar by relatively orthodox means.

The dials of Mercury, Saturn, and of the Moon were more complicated. The Mercury dial indicated three motions and consequently required elliptical wheels to

movable feasts. In leap years, two letters are required, one to February 29, and the letter next proceeding for the remainder of the year.

[27] *Indiction:* The Roman indiction derives from an edict of Emperor Constantine in A.D. 312 which provided for the assessment of a property tax at the beginning of each 15-year cycle. It continues to be used in certain ecclesiastical contracts.

accomplish them. Since the dial of the Moon had to move in increasing arcs through similar angular progression, oval wheels were employed. These were divided into sectors of unequal circumferences, but each with the same number of teeth. The Saturn dial indicated two motions of the planet.

The Primum Mobile was a sidereal dial at the front of the mechanism which incorporated the dial of the sun and was driven from the 24-hour mean dial. It completed 366 days during a period of 365 days. The Primum Mobile drove the dial of the Moon on the opposite side of the framework.

Although no mention of the materials of which the mechanism was constructed appeared in the manuscripts, contemporary accounts reported that it was constructed entirely of brass and bronze or copper. The dimensions of the astrarium as interpreted from surviving manuscript descriptions, and applied in the model constructed for the Smithsonian Institution, Washington, are 4 feet 4 inches in overall height and 30 inches in diameter at the widest point. The model consists of 297 parts, of which 107 are wheels and pinions.

NAMING OF THE MASTERPIECE

The masterpiece of Giovanni de' Dondi has been variously called "astrarium" and "planetarium" and it has been more often designated simply as a "clock." Francesco Petrarch, contemporary and intimate friend of de' Dondi, objected to the latter term, and in his last will and testament he referred to it as a planetarium:

Master John de Dundis, natural philosopher and easily the leader of astronomers, called "of the clock" on account

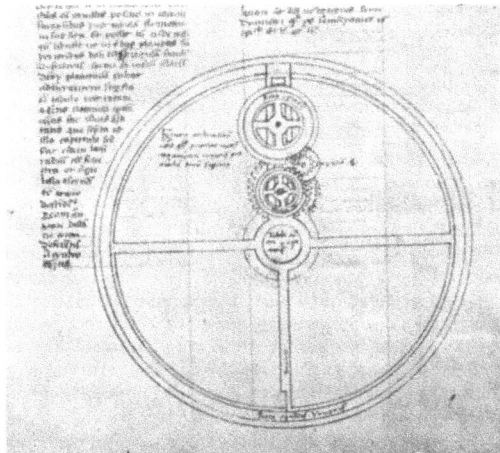

FIG. 18. Drawing, 1461, behind-the-dial work of the dial of Mars of the astrarium. From MS. Laud Misc. 620, folio 88ʳ, in the Bodleian Library, Oxford. Photograph courtesy the Curators of the Bodleian Library.

of that admirable work of the planetarium which he made, which the uneducated people think is a clock. . . .[28]

It is apparent from Petrarch's reference that the name "planetarium" was applied to the mechanism in de' Dondi's own time. If it is assumed that the two previously noted manuscripts in his private library were copies of his treatise on the astrarium, then both names for the mechanism may have been used interchangeably by de' Dondi himself. Inasmuch as the *Tractatus astrarii* referred to the planets as "the wandering stars", however, it seems likely that "astrarium" was the name first applied to the mechanism, and it has been adopted for the purpose of this study.

In his treatise on the astrarium, Giovanni de' Dondi formulated his reasons for having attempted its design and construction. He stated that he undertook the project with the hope of bringing common appreciation to the noble study of astronomy, which had been troubled and weakened by astrological fallacies, rendering many of the early studies on the movements of planets absurd. Accordingly, he hoped to construct a mechanism which could demonstrate the planetary movements constantly, without the need of additional calculations. In this enterprise he was guided by the *Theorica planetarum* of Campanus of Novara. By means of his mechanism, de' Dondi hoped to demonstrate that Aristotle had not erred in his description of the complex movements of the planets.

The passage in de' Dondi's *Tractatus astrarii* relating to the inspiration he received from Campanus of Novara is

Sumpsi autem huius propositi et ymaginationis exordium ex subtili et artificiosa ymaginatione Campani in compositione instrumentorum adequationis quam docuit in sua theorica planetarum. . . .

A translation of this passage follows:

I derived the first notion of this project and invention from the subtle and ingenious idea propounded by Campanus in his construction of equatoria, which he taught in his *Theorica planetarum*. . . .

The well-known work of Campanus of Novara, *Theorica planetarum,* was produced during the second half of the thirteenth century. In addition to the discussion of the planetary theory from which it derived its name, this work was one of the earliest and most important works on equatoria in medieval Europe. Inserted in the text were descriptions and specific instructions for the construction and use of a series of equatoria for the various planets. These data were subsequently abstracted by John of Gmunden in 1428 to form his work entitled *Compositio et usus instrumenti quod magister Campanus in Theorica sua docuit fabricare.*

It seems quite likely that in addition to consulting the *Theorica planetarum* for Campanus' theory of the

[28] Squarzafichus [68], p. 73, Document II; and Baldelli [7], Parte 2, pp. 180–195.

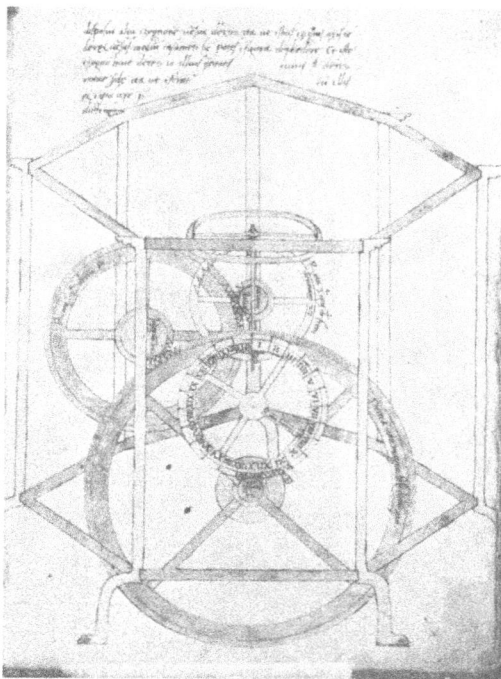

FIG. 19. Main train and horary dial of the astrarium, from MS. 248, late fifteenth century, in the Wellcome Historical Medical Library, London. Photograph courtesy the Wellcome Library.

planets, de' Dondi was inspired by it to construct a clockwork equatorium, and that his astrarium was, in fact, an equatorium with an astrolabe and calendar. Just as in the equatorium described by Campanus, where a separate plate serves for each planet, on de' Dondi's astrarium the planets are shown on individual dials instead of being combined into a single instrument as, for instance, in the equatorium described by Franciscus Sarzosius in a work published in 1526.[29]

De' Dondi certainly deviated considerably from the actual design of the planetary instruments which combined to make up Campanus' equatorium. Nevertheless, it must be stressed that when de' Dondi mentioned the *Theorica planetarum,* in his *Tractatus astrarii,* he was referring not only to a planetary theory which inspired him to produce a work of his own, but also to the discussion of an actual instrument. Consequently, de' Dondi's astrarium is not only an instrument of the

[29] *Franciscus Sarzosi Cellani Aragonei in aequatorem planetarum libri duo* [65]. An anonymous example of this type of equatorium, perhaps made by François Fine, and dating from *ca.* 1500, is preserved at the Museum of the History of Science at Oxford, No. 57–84/176. See Emmanuel Poulle and Francis Maddison [105].

greatest significance in the history of horology, but it constitutes at the same time a most important development in the history of equatoria and other astronomical models. Furthermore, the astrarium of de' Dondi is the earliest known attempt to make a mechanical equatorium.[30]

THE ASTRONOMICAL CLOCK OF JACOPO DE' DONDI

No study of the astrarium would be complete without some consideration of the possible contribution made to its design or construction by Giovanni's father, Jacopo de' Dondi.

Jacopo de' Dondi, one of the foremost physicians of his time, was born in Padua *ca.* 1290, the son of Isaaco, also a physician. After studying medicine at the University of Padua, he accepted an invitation to become municipal physician at Chioggia in 1313. The esteem in which he was held led to his being given Venetian citizenship in about 1333. He returned to Padua in about 1342. During this period he was referred to variously as *medico* (doctor) and *fisico* (physician) in contemporary documents. In addition to lecturing at the school of medicine at the University, Jacopo succeeded in extracting salt from the waters of Abano, and he experimented with its domestic and medicinal uses. This led to the writing of his compendium of theoretical and practical medical doctrine entitled *Aggregator Iacobi Dondi patavini exc. philosophi et medici liber . . .* which brought him considerable renown.

On 20 August 1355, the Prince of Carrara awarded Jacopo an exclusive privilege to extract salt and to sell it freely without taxation. However, his subsequent experimentation in this field led to problems with invidious competitors. In justification of his endeavors he wrote *Consideratio Iacobi de Dondis de causa salsedinis acquarum . . .* which later became part of the work entitled *De balneis,* published in Venice in 1553.

Equally important with his work on salt extraction and medicine, was the astronomical tower clock which it is claimed he designed and had constructed for Prince Ubertino of Carrara in March, 1344, on the tower of the Palazzo Capitanato in Padua, called the Torre dei Signori.

The earliest accounts related that the clock face included a 24-hour chapter ring, a calendar dial with lunar phases, the zodiacal signs, and that it was equipped with striking work. According to Amati,[31] writing in 1828, the clock was constructed and installed by a young

FIG. 20. The dial of the tower clock in the Piazza dei Signori, Padua. Photograph courtesy Fratelli Alinari, Florence.

Paduan named Antonio, from the designs made by and under the supervision of Jacopo de' Dondi.

The astronomical tower clock of Jacopo de' Dondi has been the subject of a spirited controversy among writers in the eighteenth and nineteenth centuries.[32] Surviving descriptions are inconsistent and conflicting. In 1560, Bernardino Scardeone[33] stated that the great

---

[30] There remains the possibility that Richard of Wallingford's clock incorporated a mechanized version of the Albion, which was basically an equatorium (see above p. 6). Further study, both of the Albion treatise and the manuscripts which apparently describe Richard's clock, will be necessary before any conclusions can be drawn on this subject.

[31] Amati [1] 2, cap. VIII: p. 166. The life and work of Jacopo de' Dondi are described in Bellemo [14], Colle [26], and the several works by Gloria [37–38–94–95].

[32] The works of these writers are described under the sections of the eighteenth and nineteenth centuries later in this study.

[33] Scardeonius [67], pp. 206–207.

clock "which is visible above the noble tower in the center of the city [Padua]" was the work of Jacopo de' Dondi. In 1589 Jacopo Cagna [34] wrote that "Giacopo (Dondi) was a doctor and astrologer, and the one who made the clock of the Palazzo dei Signori.—This was the reason that the Dondi [family] were called 'Horologi [of the Clock].'"

Later, in 1623, Angelo Pordenari noted that

Over this gate [of the Palazzo Capitanato] was an extremely beautiful tower [the roof of which was] covered with lead, on which was that most ingenious clock, which in addition to striking and indicating the hours, indicated the days of the month, the course of the sun through the twelve signs of the zodiac, the days of the moon, the aspects of the same with the sun, and its phases. The inventor of this marvelous work was Giacomo Dondi, Paduan noble and most celebrated doctor and astrologer, whose family thereafter because of this famous clock began to be called "Horologia" [of the Clock].[35]

The tower, and the clock, were subsequently destroyed in 1390 in a battle which terminated the reign of the princes of Carrara. At the request of the populace, the city council in 1423 approved the erection of a clock on the façade of the Capitanato palace. In 1428 the council accepted a design for a striking clock submitted by Novello dall'Orologio, a descendant of the de' Dondi family through another branch. In 1430 the council assigned the execution of the clock to Giovanni dalle Caldaje and in 1434 the clock was completed. This clock has survived, with modifications, to the present time. In 1688 it was fitted with a pendulum escapement by Giovanni Carleschi.[36]

There is some reason to believe that Novello designed only a movement, duplicating the original design of Jacopo and that it was deliberately planned to be a replacement. The present dial may be the original dial produced and installed by Jacopo de' Dondi in 1344.

Some confirmation of the claim that Jacopo was the inventor and designer of an astronomical clock for the tower of the Capitanato is derived from the wording of his epitaph [37] a translation of which follows:

I, Jacobus, was born in Padua, and having crept back to the earth whence I came, this confined urn conceals my cold ashes. My work was useful enough to my country, and known to my city. My art was medicine, and to know the sky and the stars, whither I now proceed, released from the prison of my body. And, truly, each art remains, adorned with my books. Yet, indeed, dear reader, know that it is my invention that, from afar, shows at the top of the lofty tower the time, and the changing hours which you count. And pray, in silence, for my peace or my pardon.

[34] Jacopo Cagna (1589), cited in Bellemo [14], pp. 197–201.
[35] Angelo Pordenari (1623), cited in Bellemo [14], pp. 197–201.
[36] Gloria, L'Orologio [37], p. 45.
[37] Illustrated and described in Gloria, I due orologi [94], frontispiece and p. 684, and examined in Padua in May, 1963, by S. A. B.

Fig. 21. Tombstone of Jacopo de' Dondi in the south wall of the baptistry of the cathedral at Padua. Photographed by S.A.B. in May 1963.

This epitaph is inscribed in a block of white marble cemented into the outer wall of the south side of the baptistry contiguous with the front of the Duomo of Padua. Although the sepulchral arches erected in memory of Jacopo and of his sons Gabriele and Giovanni were demolished and lost when the front of the cathedral was modernized, Jacopo's sepulchral stone has survived to the present. However, the dates of the inscription and of the stone are not certain. Jacopo died in 1359, between 29 April and 26 May. The earliest known published mention of the epitaph is that by Scardeone's in 1560.[38]

For a proper appraisal of the clock of Jacopo de' Dondi, a comprehensive study must be made of contemporary documents surviving in local archives and of the subsequent writings on the subject. At the present time there is no certainty regarding the design or characteristics of this timepiece; yet there is little doubt that he did indeed design or construct a public clock, probably astronomical, which was installed on the Palazzo Capitanato in Padua before the construction of the astrarium by Giovanni. If Jacopo's clock was in fact an astronomical clock, as contemporary writings indicate, then there is reason to believe that he exerted some influence on the astrarium produced by his son. Jacopo, a widower, lived in the house of his son Giovanni, from 1348 until he died in 1359. It seems unlikely that Jacopo, whose interest in the study of astronomy is well established, lived in his son's house for eleven years while the design and construction of the astrarium was in progress, without having interested himself in the project and without having made substantial contributions to it. One might even be led to speculate whether the astrarium may not, in fact, have been conceived, designed and constructed in part by Jacopo, and completed after his death by his son, Giovanni.

[38] Scardeonius, op. cit. [67], pp. 206–207.

Verification of Jacopo's interest in astronomical studies may be found in a work by Prosdocimo de' Baldomandi of Padua, *Canones de motibus corporum supercoelestium*, written about 1424, in which he stated that

The tables of planetary motion of Jacobus de Dondis of Padua, extracted from the Alfonsine tables, are easier and more convenient to use than the Alfonsine tables, and equally well, perhaps better, verified and corrected. . . .[39]

Michele Savonarola, writing in 1440 [66] attributed the corrected astronomical tables to Jacopo's son, Gabriele. However, there does not appear to be sufficient evidence to question de' Baldomandi's attribution of this work to Jacopo de' Dondi.

Jacopo's astronomical tables, as well as several others of his manuscripts, were apparently acquired after his death by Giovanni de' Dondi. The inventory which Giovanni's widow, Caterina, filed in 1389 after his death listed the following manuscripts, among others: [40]

Item 73. *Liber Ugucionis de vocabulis*
Item 101. *agregator Jacobi de Dondis*
Item 106. *Tabule Alfogi in astrologia.*

Although the extent of Jacopo de' Dondi's contribution to the production of the astrarium cannot be established on the basis of evidence now available, it is an aspect of the astrarium's history which cannot be ignored, and a question which will, hopefully, be resolved by future research.

In passages quoted above, Cagna and Pordenari claimed that the title "dall' Orologio" was made part of the family name because of the clock which Jacopo de' Dondi designed for the Palazzo Capitanato. This claim was disputed several centuries later by a descendant, Monsignor Francesco Scipione, Marchese Dondi dall'Orologio. It was his contention that the title was awarded not to Jacopo, but to his son, Giovanni de' Dondi, for his invention and construction of the astrarium. On the other hand, Gloria noted and

described documents dated before the completion of the astrarium which mention Jacopo Dondi "dall' Orologio" seem to provide conclusive evidence that the title was added to the family name after the erection of Jacopo's astronomical clock in Padua, and before the completion of the astrarium by his son.

FOURTEENTH–CENTURY RECORDS

Impressive as the astrarium appears in modern times, in the fourteenth century it was considered to be a work of ingenuity and complication beyond comprehension.

Approximately two decades after its completion, the astrarium was acquired from de' Dondi by Duke Gian Galeazzo Visconti, Conte de Virtù, and installed in the library of his great Castello Visconteo in Pavia. This castle, which survives, was constructed between 1360 and 1365, almost coinciding with the production of the astronomical clockwork. It was an imposing and majestic structure, designed on a quadrilateral plan with four great towers punctuating each corner. Each of the towers was allocated to a specific purpose by the Duke. The ground floor of the tower at the left of the entrance was divided into offices where the Duke held conferences and attended to affairs of state.

On the upper floor of this left tower was situated the famous ducal library, where Francesco Petrarch and Leonardo da Vinci came to study. It was a colorful room, perfectly square in plan with a high ceiling. The room was illuminated by high mullioned windows of two and four lights set in the center of two walls. Window seats within embrasures were approached by several steps. The ceiling and walls of the library were decorated with frescoes, and light from the windows picked out the long arrays of illuminated manuscripts bound in colored leathers, velvets and brocades which were attached by means of silver chains to shelves which lined the walls. Many of the covers bore the arms and initials of the Duke who had acquired them. Ornamenting the library, according to d'Adda [41] were

---

[39] See Bellemo [14], pp. 80–81.
[40] Lazzarini, *Libri* [96], *op. cit.*, pp. 26–32.

[41] D'Adda, *Indagini* [29], Parte Prima, pp. lxv, lxvi, 102, 108.

FIG. 22. Castello Visconteo at Pavia. The ducal library was on the second floor of the tower at the left of the drawbridge on entering. Photograph courtesy Fratelli Alinari, Florence.

two terrestrial (and celestial?) globes and an armillary sphere of gilt bronze.

As closely as can be determined,[42] the astrarium was acquired from de' Dondi by the Duke Visconti in the year 1381. Caffi [43] noted that de' Dondi presented the mechanism to the Duke as a gift in repayment for favors already received. De' Dondi first returned to Pavia in 1378, having lectured at the Università in 1372 for a brief period. In that year (1378) he was summoned to Pavia to attend Azzone, the ailing young son of the Duke. De' Dondi wrote that

When the naturally brilliant son of the magnificent and powerful lord, Duke Galeazzo Visconti of Milan and Count of Virtù, was stricken with an obstinate and grave disease, I was forced to remain with him a whole year in the city of Pavia.[44]

Note should be made of the fact that although the astrarium was installed in the ducal library in the left tower of the castle in 1381, Bossi [45] claimed that it was situated in the right front tower. This statement, which was quite in error, appears to have been based on a similar statement in an earlier work of Stefano Breventano,[46] published in 1570. It was not until considerably later, in the fifteenth century, that the library was moved to the ground floor of the right front tower.

The presence of the astrarium in the ducal library, where it was given pride of place in the center of the great room, was noted by several writers. Umberto Decembrio in a work addressed to Filippo Maria Visconti, referred to

Giovanni di Orologio of Padua, the wisest astrologer of his age, to whom, most invincible Lord [Filippo Maria Visconti], that clock which is in your very famous library gave the last name.[47]

The earliest description which provides any detail, however, is the *Songe du vieil Pelerin* written in about 1385 by Philippe de Maisieres, de' Dondi's contemporary and personal friend:

It is known that there lives in Italy at the present time a man learned in philosophy, in medicine, and astronomy to a singular and solemn degree, by general repute, excellent in all three sciences, of the city of Padua. His surname is lost: he is known as "Master John of the Clocks," who now lives with the Counts of Virtù, who for the three sciences receives each year gains and benefits in the amount of about 2000 florins or thereabouts. This Master John of the Clocks has produced, in his time, great works on the three sciences mentioned above, which by the scholars of Italy, Germany and Hungary are warranted and of great reputation: among these works he has made an instrument, by

some [persons] called a Sphere or a clock for the celestial movements: which instrument shows all the movements of the signs and of the planets with their circles and epicycles, and differences by multiplication, wheels without number, with all their divisions [teeth?], and each planet of the said sphere is shown separately. By night one sees clearly in which sign and to what degree the planets and the stars appear in the sky: this sphere is made so subtly that in spite of the multitude of wheels, which could not be numbered accurately without dismantling the instrument, the entire movement of which is governed by a single weight, which is such a great marvel that the solemn astronomers come from far regions to see in great reverence the said Master John and the work of his hands; and it is stated by all the great scholars of astronomy, philosophy and of medicine that there is no written or other human record that in this world there ever was made such a subtle or so solemn an instrument of the celestial movements, as the aforesaid clock; the subtle skill of the aforesaid Master John, who with his own hand forged the said clock, all of brass and copper, without assistance from any other person, and did nothing else for sixteen years, according to the information given to the writer of this book, who has had a great friendship with the said Master John.[48]

A longer description of the astrarium was provided in a letter written on 11 July, 1388, to de' Dondi by his friend, Giovanni Manzini of Pavia:

I saw again the globe-clock [a clock of the celestial sphere] which you made with your hands and which you brought out to take shape from the deepest recesses of your mind: to me it is a magnificent work, a work of divine speculation, a work unattainable by human genius and never produced in generations past; although Cicero tells how Posidonius had constructed a sphere which revolved, showing, through the sun, moon and five planets, what happens in the heavens at night and during the day. I do not believe that there was such competency in art at that time, nor was there such mastery of skill as is shown in this. I do not believe that any of posterity can make it or excel it, since in the passing of time we do not see such sublime growth of genius. . . . For in your work, full of artifice, worked on and perfected by your hands and carved with a skill never attained by the expert hand of any craftsman, one can see at their correct distances the distinct globes of the seven stars moving in circular orbit and called "wanderers" because of their motion; the highest of them is Saturn, whom they say seems smallest in the sky but which has the widest orbit: in the thirtieth year it slowly returns to the point from which it started and it contributes to a cold, rigid bearing. Nearby, Jupiter revolves in its orbit, only faster, completing its course in twelve years. Then the third sphere, Mars, which travels through its orbit in two years; between the warmth of Mars which is spread before it and the chill of the old man pictured in ancient times with the bent scythe, which stands above it in the first position, is the moderate Jupiter, who intervenes between them and who is known to have salutary power. And the sun, the greatest giver of light, and after omnipotent God, the only producer and propagator of all things productive, holding the mean of the most lucid and spacious celestial motions, completes its course with its annual journey through the Zodiac: and you have represented its rising and setting, the change of the seasons, the order of the months, the correspondence of the signs of the Zodiac, the lengthening and diminishing of days, the length and minutes of the hours and all the efficacy of this star with the most delicate

[42] Bossi, *Storia Pavese, ca.* 1381, cited in del' Acqua, *Palazzo ducale* [88] 1, parte, p. 44.
[43] Caffi, *Castello* [85], pp. 543–558.
[44] Giovanni de' Dondi, *De Balneis* [32]. See reference to "De fontibus calidis agri Patavini" (Document III).
[45] Bossi, *Storia Pavese* [88]. See n. 42.
[46] Breventano [19], Lib. 1, folio 7 (Document XXX).
[47] Ubertus Decembrius cited by Magenta [48], p. 219 fin. (Document IV).

[48] De Maisieres [90], Tome XVI, pp. 227–228 (Document V).

delineations of designs in single positions and points—all this you have shown with greatest acuteness. The planet Venus follows, traveling through space on an alternate path; for preceding sunrise, that noble harbinger, first named Lucifer, acquired the name Hesperus from that part of the West prolonging the light. The influence of this planet generates everything on earth. For according to Pliny Veronese, spreading fecund dew in orchards, it not only fulfills the conception of the earth but it also stimulates the regeneration of all the animals. It travels on its orbit in as many days as Phoebus, never wandering more than fifty-six degrees from it, as Timaeus teaches. In the next to last place, going down, you placed Mercury, next to Venus but not of equal effectiveness or greatness. This one completes its orbit nine days sooner; now shining before sunrise, now after sunset, it never loses more than twenty-two degrees, as Sosigenes taught. As last of the planets you present Phoebe, nearest earth, who was the Moon, Diana, Dittina, Proserpine, and Cinzia to our ancient poets, named after various phases; to diminish nightly darkness she appears in several phases, now thin and spare, now full and well-rounded, now curved, now roundish, now resplendent with youth, now languid with age, now the reddening bearer of winds, now the pale harbinger of rain, now the conciliator of tranquil weather with the lustre of pure serenity of which we, who live in this low world, are so often deprived; now she hides, now she appears later, now sooner, now she is encircled with a halo now clear. now rosy; and sometimes she seems sanguine, about to announce great calamities; sometimes she rises highest in the expanses of the sky, sometimes she touches mountains; sometimes she rises in the North; sometimes, she sets in the south; and her orbit is the smallest of them all, for in twenty-seven and one-third days she makes the trip through the sky which the cold orb of Saturn completes in thirty years. She divests her brother of the splendor of his rays and dresses again, waxes and wanes, and one knows that through the intervention of the moon the sun is hidden and through the opposition of the earth the moon is hidden, for the sun's rays being impeded, the moon is hidden by the opposition of the earth and sudden darkness steals the brilliance of the room through the interposition of the earth. You have represented all these things clearly to him who attentively watches, with living, effective images. But I, ignorant of these things, have talked enough of astronomy, about which I would have been silent with others, and even with you, if my confident boldness inspired by you had not been free and ready; however, I would not be so bold as to allow my discourse to fall in someone else's hands, for I could easily be derided. . . . Finally, there is that lower ring found in your work where the days of each month are seen written in order, the reckoning of the whole year, and the calendar, and where there are certain little windows hollowed out with recesses which indicate the festive solemnity occurring to be celebrated that day, and which reveals the name of the saint written circularly, causing no small wonder to the ignorant and great delight to those who understand it. In short, I conclude that there was never invented an artifice so excellent and marvelous and of such supreme genius.[49]

Clearly, the astrarium was highly prized and every effort was probably exerted to preserve it and maintain it in operating condition. Of unique interest in this regard is an entry in the register of bills, receipts and

[49] Lazzari, *Miscellaneorum* [41], Tome I, pp. 173, 226 (Document VI).

FIG. 23. Coat-of-arms of the Visconti-Sforza rulers of Milan. From a stone carving in the Certosa of Pavia. Photograph courtesy Fratelli Alinari, Florence.

monthly salaries charged to the city of Pavia which was compiled in 1399:

Giovanni de Clarii master of clocks [master clockmaker] assigned to care for the clock in the great castle of Pavia for his monthly salary L. 6. 8.[50]

It is apparent from this entry that the master clockmaker, Giovanni de Clarii, was hired not to repair the astrarium, but to maintain it.

FIFTEENTH–CENTURY HISTORY

The fame of the astrarium was not diminished by the passage of the years. In his biography of Duke Filippo Maria Visconti, Pier Candido Decembrio noted that he

[50] *Registro delle bollette . . . Pavia* [137], 1399, folio 56 (Document VII).

... had a famous clock in the library of Pavia, memorable above all those of our time and almost divine, made by Giovanni of Padua, the distinguished astronomer. . . .[51]

There has been some speculation as to whether the astrarium as reproduced in the models at the Smithsonian Institution and at the Museo Nazionale in Milan is complete, or whether the original was furnished with a dust cover. In an article on the subject of the mechanism Captain Antonio Simoni had shown it with a cover over the top of a Brunelleschian design.[52] The problem is resolved in a letter addressed to the Duke from his chancellor, Facino de Fabriano, dated from Pavia 17 April 1456. After describing various needs for the ducal library, and the inventory of the volumes, which was in progress, he reported that

... [And other than this] It is necessary to have repaired the cupboards which are being ruined, and to repair several benches used for sitting by those who wish to study and which are too narrow and difficult to enter, but this is a

small matter: but it would be a wise thing to undertake to repair the cover of the *Astrolabio* so that it could be lifted up cleverly by means of some pulley, that being as it is, if the greatest care is not taken, [then] it falls upon the books and ruins the cover and the books, and this would be to the cost of your Imperial books. . . .[53]

This letter forces a revision of the modern concept of the appearance of the astrarium. It is apparent from the foregoing that the mechanism was protected from dust by a cover, which was extremely heavy. It is possible to visualize only a cumbersome construction, made probably of bronze or gilded metal to harmonize with the movement of the astrarium, which may have served as a frame for the insertion of glass panels to permit viewing of the dials. It is possible that the cover might have had doors or panels which opened to reveal the dials of the clock, instead of fixed glass panels. The cover may also have been fitted closely to the clock frame to exclude dust from the mechanism leaving the dials uncovered. It is not perhaps unreasonable to assume a certain resemblance between the general

[51] Decembrio, *Vita* [31], Tome XX; p. 1017 (Document VIII).

[52] Simoni [114], pp. 7-16.

[53] *Missive* [119] (Document IX).

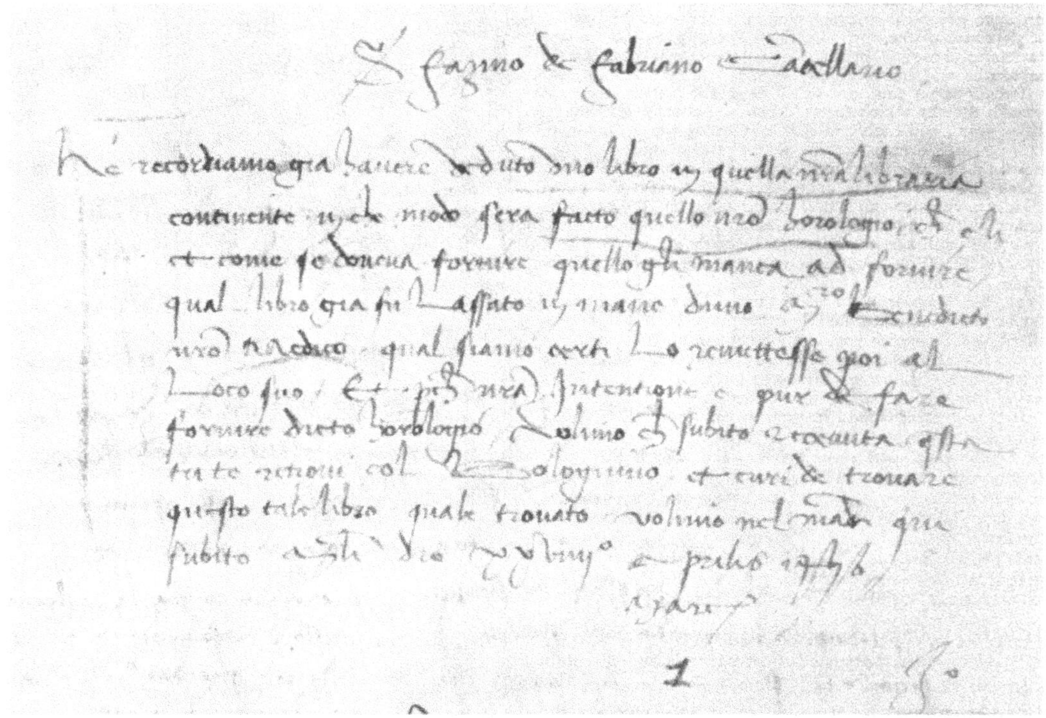

FIG. 24. Letter from Duke Francesco Sforza to his chancellor, Facino da Fabriano dated 29 April 1456, requesting the book describing "our clock." Archivio di Stato, Milan. *Missive Ducali Registro No. 32.* Photograph courtesy Archivio di Stato, Milan.

outline of the cover of the astrarium, and the clock tower or frame illustrated in the sketchbook of Villard de Honnecourt, *ca.* 1235,[54] as well as the frame of the clock in a fifteenth-century miniature painting illustrating a French translation of Heinrich Suso's *Horloge de Sapience* [54a] in the Bibliothèque Nationale in Paris.

It is to be noted that in this letter Facino referred to the mechanism as an "astrolabio," a name which is not accurate but which occurred repeatedly in subsequent records. The confusion may have arisen because the astrolabe was a well-known instrument for solving problems relating to the position of the sun and the "fixed" stars, and its name is similar to the word astrarium which refers to the "wandering stars" or planets. Furthermore, the dial of the primum mobile on the astrarium includes some essentials of an astrolabe—the zodiacal circle moving over a stereographic projection of the horizon and almuncantars (circles of altitude), and lines of unequal hours.

The astrarium appears to have been the subject of some Ducal preoccupation, for less than two weeks later the Duke communicated with Facino da Fabriano in a letter dated 29 April 1456, from Milan, saying that

We remember having seen a book there in our library containing information about the manner in which was constructed our clock which is there, and how that which is missing may be supplied, which volume had been left in the hands of a Master Benedict, our physician, whom we are certain returned it to its proper place. And since it is our intention to restore the aforesaid clock, we desire that as soon as you have received this you get together with Bolognino and attempt to find this aforesaid volume which as soon as you have found it you are to send me at the earliest opportunity.[55]

Facino replied promptly in a letter dated 1 May 1456, in which he informed the Duke about the inventory of the library's contents then in progress and other matters. He went on to state that

. . . Concerning the astrolabe or indeed the clock [*lastrolobio ovvero arrologio*] I believe that Your Highness will wish to proceed as is necessary : but I do not know whether there are in places other than Milan any master [clockmakers] ; and although there are no others that I know about there in Milan, there is someone here for this work. I do not say that he understands the motions and the courses of the planets : but for the other things he has the disposition [ability] to do them as no other has, and thus I believe it would serve as an experiment. We could see, provided that it meets with the approval of Your Highness, and I believe that the worker would improve the astrolobio sufficiently by actually working it out and not from thinking it out theoretically, and this experiment that he attempts will be such that Your Excellency would not have believed it possible that he did not approach it with a theoretical understanding.

[54] See fig. 25. See also Bowie (ed.), *The Sketchbook of Villard de Honnecourt* [18], p. 66.
[54a] On Suso's *Horloge de Sapience* and illustrations of clocks in the various manuscripts, see Spencer, "L'Horloge de Sapience" [116a] and Michel, "L'Horloge de Sapience" [100] ; the former article contains an extensive bibliography.
[55] *Missive* [120] (Document X).

Recently I wrote to Your Highness about various items that are needed in this library, some required for decoration, some out of necessity and still others for convenience, to which I have never had any reply : and since you did not write, I then asked you in a note sent with don Bartholomeo Trovamala. . . . Once again there are the cupboards that tumble down, the cover of the Astrolabio which in the removal and replacement and leaving off damages every thing of the benches for sitting. . . .[56]

Subsequently the astrarium fell into disrepair, and during the fifteenth century considerable difficulty was experienced in maintaining it in working order. This is reported particularly in an account written by Michele Savonarola, grandfather of the famous Fra Girolamo :

Nor was the faithful House of de Dondis, named ab Horologio (about which later), of our city, of small moment in this category of writing. This house had so many famous men of medicine, whose teaching and fame still exist. At the beginning I take up in third place Joannes ab Horologio, a man, as you know, almost divine. He was almighty in the teaching of medicine. He was a great orator, a practicing doctor, the greatest mathema-

[56] *Lettere* [121] (Document XI).

FIG. 25. Drawing of a clock frame or tower, from the sketchbook of Villard d'Honnecourt, *ca.* 1235, Bibliothèque Nationale, Paris, *MS français*, 19093, p. 12.

FIG. 26. Miniature painting in a fifteenth-century manuscript of the *Horloge de Sapience*, a French translation of a work by the fourteenth-century mystic, Heinrich Suso, Bibliothèque Nationale, Paris, MS. 43657, Fr. 455, folio 9. Photograph courtesy Bibliothèque Nationale, Paris.

tician, an admirable craftsman with his hands, whose wisdom, learning and genius were observed by Francesco Petrarca, who, in one of his letters lists the marvelous gifts of this admirable man and claims that no one in the world is a more learned man; what detracts from all these—he worked servilely.

He constructed at Pavia a clock by his hands and genius; its beauty must be admired, in which are the firmament and the spheres of all the planets, so that thus might be included the motions of all the stars as in heaven.

It showed the decreed feasts on the days and many other things wonderful to behold. Such was the wonderful assembly of his clock that after his death no astrologer knew how to correct it and equip it with proper weights. Only recently an astrologer and a great craftsman, drawn by the fame of such a clock, came from France to Pavia and spent many days in collecting together the wheels. And at last it happened that he put it together in that one order which was proper and give it the motion which it should have. Indeed I think, my Anthony, that it ought to be included and placed among the wonders of the world: it is a thing which it is amazing to hear about and which was never before heard about in the world nor was such a unique thing seen. Every other thing has its equal or almost its equal.

From this clock the glorious house of de Dondi received its cognomen. The bones of such a man are held in a casket which is elevated and greatly decorated at the first door of the cathedral church.[57]

A puzzling remark in Savonarola's account is the mention of the arrival of a craftsman from France who successfully repaired the astrarium. This took place "only recently," and so occurred not long before 1440, when Savonarola's work was written. It is particularly of interest because sixteen years later, in 1456, the Duke requested the services of a clockmaker at the French court in Paris to come to Pavia to repair the astrarium once more. The request was obviously for the services of the same craftsman who had repaired the astrarium on the earlier occasion. The request was formulated in a letter,[58] dated 12 May 1456, from the Duke's secretary in Milan to a certain Antonio de Tritio, who was probably the Duke's representative at the French court:

Antonio de Tritio.
Because we want the clock in the library of our castle in Pavia to be attended by the hands of the Master Guglielmo of Paris, who is there in the service of His Majesty, the Most Serene King, because he is a good master craftsman and at another time, within the good memory of the late, most Illustrious lord, he had it in his hands and had begun to repair it, and because it really is a marvelous work: consequently I want to beg His Majesty the King that he deem it worthy to provide the services of the said master for at least four months, and if we are able to have him, I wish him sent here to us as soon as possible. Given in Milan the 12 May 1456.

Cristof.
C.

Efforts to identify a clockmaker in the service of the French court at this period have proven incon-clusive. Later writers, including Cardano,[59] Tassoni,[60] and Montucla,[61] have identified him as Guglielmo Zelandino, but all supporting references to him as an Italian clockmaker have led to further confusion. He allegedly worked in the first half of the fourteenth century and was reported to have constructed the turret clock on the campanile of the church of San Gottardo in Milan in 1335. Other sources relate that he worked in the second half of the fifteenth century and that a clock which he had made in 1402 had been offered to the Emperor Charles V by the city of Paris on the occasion of his coronation in 1529.

It seems far more likely that the clockmaker Guglielmo Zelandino may be identified with Guillelmus Zelandinus (William of Zealand) also known as Willem Gilliszoon of Carpentras, and Willelmus Aegidii de Wissekerke. Although a native of Zealand, he settled at Carpentras in southern France. He first comes to our attention in a number of accounts of the Sicilian King René between 1476 and 1480, during which period he is stated to have furnished King René with a variety of gnomonic and astronomical instruments, as well as two small clocks with weights, a round clock, and astrolabes. A biographical note by Simon of Phares in 1495 reported that he made spheres for the King of Sicily and for the Duke of Milan. One sphere that he made for King René cost 1,200 escuz. This contained numerous complications and was constructed in such a manner that it exhibited all the movements of the planets at all hours of the day and night. Guillelmus Zelandinus was the author of a treatise on an equatorium written at Carpentras in 1494, of which a manuscript and printed versions survive, and of an astronomical work known from a manuscript.[62] In considering the dates known for Guillelmus Zelandinus, from 1476 to 1495, it seems entirely feasible that it was indeed he who repaired the astrarium on both occasions, just before 1440 and again in 1456.

The possible association of Guillelmus Zelandinus with the astrarium at Pavia, and his known association with King René, lead to an interesting speculation. In September 1453, King René visited Pavia, where he was received by Duke Francesco Sforza with great honor. After touring the city, René visited the library in the Castello Visconteo where he presumably saw the astrarium, and the Duke honored him by presenting him with keys to the castle. Inasmuch as Guillelmus Zelandinus probably repaired the astrarium just before 1440 and again in 1456, it is possible that his skill was mentioned to René during the latter's visit to Pavia.

[57] Savonarola, *Commentariolus* [66], Tome XXIV, p. 1164 (Document XII).
[58] *Missive Ducale* [122] (Document XIII).

[59] Cardanus, *De Subtilitate* [22], Libro XVII, "De artibus," pp. 611–612. See n. 100.
[60] Tassoni, *Pensieri*, Lib. X, Cap. XXIII, cited by Gimma [36], 1, Cap. XV: p. 127.
[61] Montucla, *Histoire* [52], Lib. I, parte III, p. 534, an VII.
[62] Thorndike, *History of Magic* [70] 4: pp. 558–560; Poulle [104].

A reference in a work by Malaguzzi Valeri is confusing:

Later, at the height of the fifteenth century, the great clock no longer functioned, and in vain the Dukes commissioned mechanicians and clockmakers—Guglielmo Zelandino, the master Claudio, a certain Zanino who had already constructed a similar instrument for the castle of Milan in 1478—to repair it. The astrolabio [astrarium] was then dismantled and removed from the library and dispatched, in 1494, to a hall in the castle of Rosate.[63]

These remarks open up new prospects. Very little is known about the master clockmaker named Claudio. It is a matter of record that for reasons not stated Master Claudio the clockmaker was banished from the Dukedom of Milan on 4 May 1475. Although it has been suggested, this clockmaker cannot be identified with Claudio, lay brother of the Dominican Order, who constructed a new clock for the church of San Eustorgio in Milan in 1572.[64]

The third name mentioned is that of the master clockmaker Zanino, who was commissioned to construct the tower clock of the Sforza castle in Milan by Francesco Sforza in 1457. According to Beltrami,[65] Zanino was commissioned in about 1478 to restore the de' Dondi astrarium which was then preserved in the castle at Pavia. Beltrami noted, however, that like so many others, Zanino did not have the courage to assume this grave responsibility.

Earlier, however, what appears to be a reference to Zanino occurs in a letter from Cicco Simonetta, secretary of Francesco Sforza, written from Milan on 28 May 1459, to Conte Bolognino de Attendolis. Simonetta wrote:

Having continually on our mind this library of ours, and deliberating on the addition of more books to it, and augmenting and maintaining all that we have had done, we have seen and well considered that if it cannot be attended diligently in a short space of time, nothing will remain of what has been done, and we are considering some qualified person capable of it. It seems that the master clockmaker who is constructing the clock there in our castle is sufficiently suitable and capable of maintaining both that clock and the one in our library, and see to it that we can provide him with a suitable salary, to which expense it seems in any manner you should contribute a portion, according to Facino da Fabriano our chancellor, whom we have taxed with it. He will tell you of it and you must believe him in this and you must give him complete faith as to our own person.[66]

Facino's mention of a clockmaker in May 1456 (see above, p. 23), seems to be related to Simonetta's reference to the clockmaker constructing a turret clock for the castle, and suggests that both communications referred to Zanino.

Another clockmaker who may have been assigned to the maintenance of the astrarium was a certain Gaspare d'Allemagna. His name appears on the rolls of the persons employed by Duke Galeazzo Maria Sforza in 1467, wherein he is designated "custos relogii" or "custodian of the clocks," of the castle at Pavia. In the rolls for 1470 Gaspare's name is absent, and he was apparently no longer employed. His name indicates that he may have been a monk from Germany, but further identification has not been possible.

The constant preoccupation of the Dukes of Milan with the maintenance and repair of their prized possession, the astrarium, is perhaps reflected again in a letter dated 6 January 1460, to Conte Bolognino de Attendolis stating that Bolognino had been informed in writing ". . . to show the clock, which is in the library and all the reliquaries, to the Master Marco de Raynis of Milan."[67]

Of considerable importance to historians of science is the record of another visitor who came to the ducal library at Pavia to examine the astrarium. This was none other than the German mathematician and astronomer, Johann Müller called Regiomontanus (1436–1476). Regiomontanus reported having seen the astrarium in 1463 and subsequently referred to it in his introductory lecture on the mathematical sciences at the University of Padua. In the course of a brief survey of the history of astronomy and the work of famous astronomers, he said:

. . . But why spin out the story? There has appeared a more recent astronomer, Johannes de Dondis, your most renowned fellow-citizen, who for so long both truly cultivated this divine subject, and pursued it in such a way, that his immortal remains [viz., his works] will be able to instruct. Really, shall you consider unworthy of notice his astrarium which is, today, kept by the Duke of Milan in the castle at Pavia? In order to see it, innumerable prelates and princes have flocked to that place as if they were about to see a miracle, and indeed not without cause, so great, and indeed so unusual, is the beauty and utility of this work, that there is no one who does not admire it. Behold, the monument of your philosopher will never perish. You, yourselves, shall be the judges whether the study of medicine bred distinction for your fellow-citizen, or whether [it was] rather [his] knowledge of the stars, for both on the one [subject] and on the other, he equally expended much labour. On these matters, an individual says nothing: they have made his name immortal to posterity. Rejoice then, O noble Paduans, to whom the studies of famous men have always been a grace. . . .[68]

[63] Malaguzzi Valeri, *Vita Privata* [49], p. 635 (Document XIV).

[64] D'Adda, *Indagini* [30], Parte I, appendice, p. 96, Doc. CXL.

[65] Beltrami, *Castello* [17]. Beltrami reported that Zanino was commissioned by Cicco Simonetta on behalf of the youthful Duke Gian Galeazzo Sforza to construct a clock for the Castello a Porta Giovia in Milan. Verification of this commission is to be found in another letter in the same archives at Milan written by the Duke from Cremona on 24 October 1457, in which he ordered Master Giacomo Imperiale and Danesio Mainerio, engineers in the Duke's employ, to come to him, together "with Master Zanino, he who is making our clock there in the castle."

[66] *Missive*, Simonetta [123] (Document XV).

[67] *Registro delle Missive* [124] (Document XVI).

[68] "Oratio introductoria in omnes scientias Mathematicas Ioannis de Regiomonte, Patauij habita, com Alfraganum publice praelegeret" in [Regiomontanus,] . . . *Rudimenta astronomica Alfragani* . . . , [62a] sig. 2ʳ⁻ᵛ (Document XVIa).

It is of interest that Regiomontanus called it an astrarium rather than a planetarium. In 1474 Regiomontanus again referred to the astrarium stating that he had such a mechanism under construction in his own workshop in Nuremberg. It was never completed in his lifetime, however, inasmuch as it was listed as incomplete in the inventory of his estate following his death in 1476.[69]

There is another context in which Regiomontanus' interest in the astrarium may be viewed. Among Regiomontanus' associates in Italy was the famous Polish astronomer Marcin Bylica of Olkusz (1433–1493), a pupil of the Jagellonian University of Cracow, who later became court astrologer to King Matthias Corvinus at Buda.[70] Bylica bequeathed to the University at Cracow his magnificent astronomical *instrumentarium*. The arrival of these instruments at the University in 1494 appears to have aroused much interest, for special arrangements [71] were made for members of the University (among whom at this time was Copernicus) to see the instruments.[72] It is perhaps not surprising that in this very year, when the arrival of these superb instruments should have excited a university which was one of the intellectual centers of Central Europe and whose astronomical school was among the first in Europe and than at its apogee,[73] a certain Henricus Ragnetensis should copy, "in studio vniuersale Cracoviensi," de' Dondi's work on the astrarium.[74] This manuscript was perhaps made for, and certainly belonged to, Leonardus a Dobczyce who flourished *ca.* 1504; it is now in the Library of the University. Another manuscript of de' Dondi's treatise also survives at Cracow.[75] This second manuscript has been considered to be of the sixteenth century, but it is more likely that it dates from the last quarter of the fifteenth century. It was written in Italy, probably at Padua or in the Veneto, in a fine humanistic cursive script, and great care was taken in the layout of the text and full series of illustrations. It

is reasonable to suggest that this manuscript is the original of Henricus' copy which, except at the beginning, has blank spaces for the illustrations, these never having been completed. One would like to know how this Italian manuscript came to Cracow. Did Bylica's bequest include a manuscript of de' Dondi's work? [76] The maker of the globe and of other instruments belonging to Bylica was Hans Dorn of Vienna (d. 1509).[77] Dorn had arrived at Buda in 1476 to continue the work on equipment for the observatory there, which had been interrupted by the departure of Regiomontanus in 1471. In 1478 Dorn was sent by King Matthias to Nuremberg to buy the books and instruments that remained there after the death of Regiomontanus.[78] All the belongings of Regiomontanus were purchased by Bernhard Walther, however, and Dorn was unable to acquire any of the instruments and books in his former colleague's library.[79] This does not necessarily imply that he returned to Buda entirely empty-handed. During the six months that Dorn spent in Nuremberg he had ample opportunity to make a leisurely examination of the contents of Regiomontanus' library. He would certainly have studied the two *Imagines celi*, which he had been particularly anxious to acquire, for instance. At the same time, Dorn would have been greatly interested in a manuscript of de' Dondi's work, had Regiomontanus been using a manuscript (which he had ample opportunity to acquire in Padua or elsewhere in Italy) while constructing his own unfinished astronomical clock. Whether or not it may be possible eventually to relate these miscellaneous facts, it is nevertheless of interest that important centers of astronomical activity, such as Nuremberg and Cracow in the fifteenth century, should have been studying de' Dondi's masterpiece.

Following upon Francesco Sforza's recollection of a volume in his Pavia library in which was explained the construction of the astrarium [80] was the subsequent discovery that a book, which is probably to be identified with it, was in the possession of the abbot of San Faustino maggiore in Brescia. On 26 September 1463, Cicco Simonetta wrote in the Duke's name to the abbot requesting its return:

I have learned from Pietro Ferrante, our godfather, of your liberality, which I wish to employ in requesting that you send us if you please the book "Astrolabio" and the quadrante [dial-plate?] which as you know we have had requested, and because we are most anxious to have the aforesaid items we are sending there to you post haste Bernardo de Corsicho, our Milanese exhibitor at present, to whom we wish that you would kindly consign and give the items stated, for which we would be most grateful and for which we offer in exchange any favor you may desire.
Milan, 26 September 1463.

　　　　　　　　　　　　　　　　　　　　　　　Cichus.[81]

---

[69] Zinner, "Planetenuhren" [118], pp. 18–20; also Zinner, *Regiomontanus* [80], pp. 82–83; 162–163.

[70] Birkenmajer, "Marcin Bylica" [84] (not available to us); Ameisenova, *The Globe of Martin Bylica* [2]; Przypkowski, "Premières cartes" [112].

[71] Copernicus was there from 1492 to 1496.

[72] Henryk Barycz, *Conclusiones Universitatis Cracoviensis ab anno 1441 ad annum 1589* (Polska Akademja Umiejętnósci. Archivum Komisji do Dziejów Oswiaty i Szkolnictwa w Polsce, no. 2, Cracow, 1933), p. 80: "Anno, quo supra [1494], die Mercurii decima septembris, facta convocacione die ad videnda instrumenta astronomica nova, missa ex Hungaria per magistrum Martinum de Ilkusch, plebanum Budensem et audiendam peticionem magistri Bernhardi . . ."; *cf.* Ameisenowa, p. 11, citing Birkenmajer: "The legacy . . . was so highly valued that on October 10, 1494, Joannes Sacranus, the then Rector of the University, excused all the students and masters from their work in order to enable them to see these beautiful and then uncommon instruments."

[73] See Barycz, *Historja Universytetu* [9]; *idem, The Development of University Education* [10]. *Cf.* also Olszewski, "Outline," p. 249.

[74] See below, p. 42, no. 8.

[75] See below, p. 42, no. 9.

[76] Bylica's will is, unfortunately, not known to have survived.

[77] Przypkowski [112], *op. cit., passim.*

[78] Ameisenowa [2], p. 58.

[79] Zinner, *Regiomontanus* [80], p. 122.

[80] See above, p. 23.

[81] *Missive* [125] (Document XVII).

One wonders whether this volume of the "Astrolabio" which described the construction of the astrarium, was not in fact a manuscript copy of the *Tractatus astrarii* which Giovanni de' Dondi may have given to Duke Gian Galeazzo in 1381 with the original astrarium. Several manuscript copies of the work must have been produced at approximately the same time, inasmuch as two manuscripts survived in the de' Dondi family, as noted by Monsignor Francesco Scipione Marchese Dondi dall' Orologio. One was the copy he subsequently donated to the Biblioteca Capitolare of Padua in 1795.[82]

The particular work describing the construction of the astrarium, was noted again in a letter from Francesco Sforza to Conte Bolognini de Attendolis dated 5 November 1464[83] in which he ordered him to consign to Franchino Cajmi, tutor of his children, who were then at Abbiategrasso, several works which would serve for their instruction. Upon departure from the Abiate the books were to be returned, and it was to be the Conte's responsibility to have the books in hand and to replace them in their proper place, and that he was

. . . returning to you the book of the clock which likewise you will have returned to its proper place. And of the replacement of this volume and of the consignment of those mentioned in the foregoing you will kindly notify me by letter.

The volume entitled "Astrolabio" is referred to again in a letter from Duke Galeazzo Maria Sforza in Vigevano to Francesco Sforza dated 14 February 1471.

We had deliberated to not permit the removal from our library in Pavia of the copy of Virgil, which Your Highness had requested to borrow for twenty days. But having enjoyed the sweet wine which you sent us, we have changed our thought, and thus wish to humor you with the aforementioned Virgil, but in addition to this we are pleased to comply also with the "Astrolabio," which is in the same library. We are accordingly writing to our custodian of the castle in Pavia for the same, who will give the Virgil to whoever is designated by Your Highness.

Vigevano, 14 February 1471.

ALEXANDER

CICHUS.[84]

There exist three inventories of the library in the Castello at Pavia. The earliest is the "Consignatio" of 1426, listing 988 items. Another inventory, made in 1450, lists only 824 items, despite some new acquisitions. A third inventory, of 1469, records 126 new manuscripts from the personal library of Duke Galeazzo Maria Sforza who deposited them in the library of the Castello in October of that year. None of these inventories mentions any work which can be identified with de' Dondi's treatise on his astrarium. This is surprising because the inventories were carefully made by highly competent persons.[85] The correspondence quoted above shows that the de' Dondi treatise was not at all times in the library, however, and it may well have been elsewhere when the inventories were made. When the manuscript was finally lost from the library, it is not possible to determine. It has been commonly accepted, following Magenta's view, that the ducal library was dispersed firstly by seizure of part of the collection of manuscripts by order of Louis XII, and secondly by the pillage of the rest by all who were able to enter the Castello during the disorders consequent upon the occupation of 1499–1550. Élizabeth Pellegrin, however, remarked that it is very probable that from 1447, after the death of Duke Filippo Maria Visconti, the various disorders which occurred favored the dispersal and the loss of some manuscripts and, in particular, of nearly the whole collection of the dead Duke.[86] Perhaps some of the correspondents quoted above were referring to a manuscript which, unknown to them, had already disappeared.

It is important to record the possible existence of another work describing the technical construction of the astrarium. According to Magenta,[87] this is a manuscript by a certain Paolo Trizio which covered the same scope as the *Tractatus astrarii* of de' Dondi. Efforts to identify this manuscript or its writer further have been unsuccessful. However, a manuscript of the late fifteenth or early sixteenth century in the Biblioteca Ambrosiana, Milan, describing the construction and use of the astrolabe, is attributed to a Paulus Tritius.[88] This author is probably the Paolo Trizio referred to by Magenta, and if he wrote on the astrolabe, then he might well have written an original description of the astrarium. Perhaps such a text survives in some library, awaiting identification. Alternatively, Paolo Trizio may merely have been the scribe of one of the copies made of the work

[82] See Manuscripts section, No. 1.

[83] *Missive* [126] (Document XVIII).

[84] Registro [127] (Document XIX). Signed by Cicco Simonetta.

[85] The inventories have been published in full, and the items recorded therein identified wherever possible, by Pellegrin, *La Bibliothèque des Visconti et des Sforza* [57c].

[86] Pellegrin [57c], pp. 70–72.

[87] Magenta, *Visconti* [48] 1: p. 569.

[88] MS. I.20.Sup. According to the Inventario Ceruti (Modern Language Association microfilm MLA 1271F, roll 3) this Latin manuscript is on paper, 20 cm. × 15 cm., in good condition and complete. The text occupies 44 folios. The manuscript, described as "Mathematicae tractatus," is said to be of the sixteenth century and to have been in the Biblioteca Ambrosiana since its foundation. Kristeller, *Iter italicum*, dating the manuscript to the fifteenth or sixteenth centuries, describes it as an "Astrological treatise, attributed by a later hand to Paulus Tritius." We have seen photocopies of parts of this manuscript and there is no doubt that the text describes the construction and use of the astrolabe and that it has no connection with the astrarium of Giovanni de' Dondi. The handwriting is consistent with a late fifteenth- or early sixteenth-century date. At the beginning of the manuscript, on the page opposite the beginning of the text, is written in a different hand: "Pauli Tritij, ni fallor, quaedam mathem. Olgiatus uidit anno 1603," and, below, in a third hand: "Felicibus auspicijs IIImi Card. Federici Borrs."

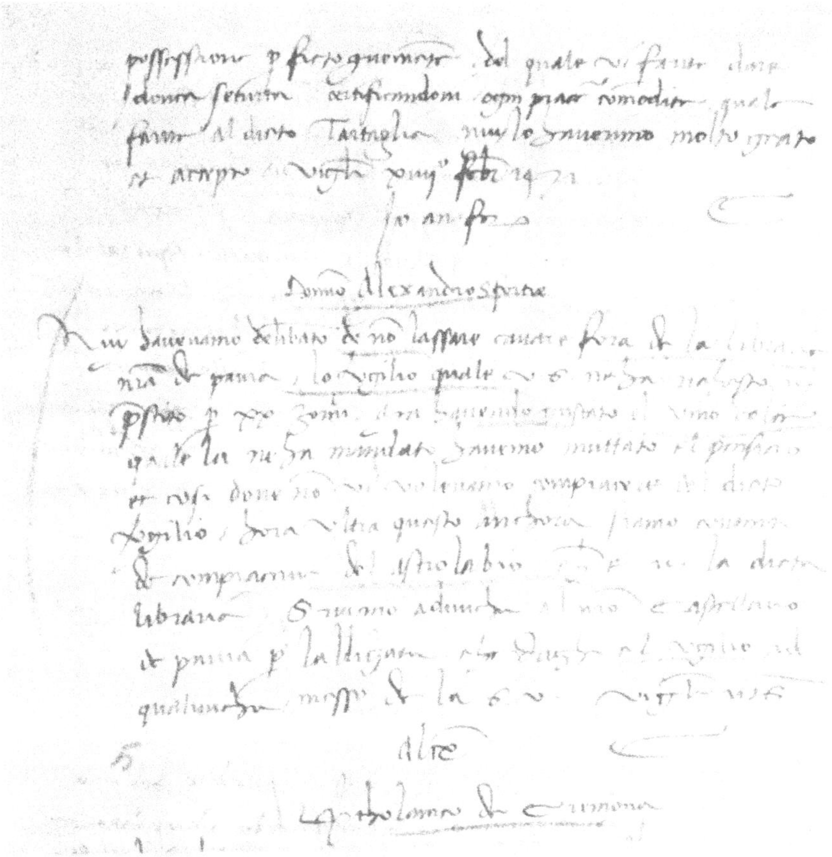

Fig. 27. Letter dated 14 February 1471, from Duke Galeazzo Maria Sforza to Francesco Sforza, informing him that he may borrow the book of the *Astrolabio* in the ducal library at Pavia. Archivio di Stato, Milan, *Missive Ducali Registro N. 98*. Photograph courtesy Archivio di Stato, Milan.

by Giovanni de' Dondi himself. It is interesting that Paolo Trizio's surname is the same as that of at least two personages associated with the rulers of Milan. A certain Antonio de Tritio for instance, was probably a representative of Duke Francesco Sforza at the French court in Paris in 1456 (see above, p. 25). A Jehan de Trecio was physician of Duke Gian Galeazzo Visconti, Conte di Virtú, and was sent by the Duke to Avignon in 1385 on a mission to the Pope.[89]

Although tradition informs us that the astrarium was the conversation piece of the times and that astronomers, scientists, and scholars from all parts of Europe visited Pavia to examine the masterpiece, no subsequent con-

temporary descriptions of it have come to light, nor were any contemporary illustrations of it known until recently.

In view of its wide renown, it is not surprising to discover that the mechanism was an object of interest for that artist, engineer, and observer, Leonardo da Vinci. A careful study of his manuscripts has revealed that Leonardo was a frequent visitor to the ducal library at Pavia during his sojourn in that city in 1489–1490. Mention of the castle library at Pavia occurs several times in his notebooks. A notation made in 1490 was a reminder to himself to ". . . try to get Vitelone which is in the library at Pavia, and which treats of mathematics." [90] Leonardo was referring to one of Witelo's

[89] Wickersheimer, *Dictionnaire biographique des médecins en France au moyen âge* [77], p. 495.

[90] Leonardo da Vinci, *Atlantico* [140], folio 225 (1490).

manuscripts on perspective, according to another note which appears in Manuscript B.[91]

During his frequent visits to the library before his departure for Milan in January 1491, Leonardo undoubtedly often admired and studied the de' Dondi clock which occupied so prominent a place in the center of the floor of the library. In this period the library was moved to the new location on the ground floor of the right tower, according to the sources noted by Malaguzzi Valeri.[92]

It is tempting to speculate whether this masterpiece of mechanism might not have provided Leonardo with the inspiration for the creation of his astronomical stage setting for the masque "Il Paradiso." Although no attempt has been made previously to link this famous stage presentation with the de' Dondi clock, the coincidence is too great not to merit careful consideration.

The masque was staged as the concluding event of the nuptial ceremonies held by Duke Lodovico il Moro to celebrate the union between his nephew, Gian Galeazzo Sforza and Isabella of Aragon on 13 January, 1490, at the castle of Porta Giovia in Milan. According to a poetical account of 1493 by Bernardino Bellincioni, the Duke's court poet,[93] the theme of "Il Paradiso" for the festival was suggested by the Duke himself. Bellincioni suggested that it was probable that the masque had originally been intended to be performed for the wedding feast of Gian Galeazzo and Isabella in the previous year, 1489, but that it was postponed by the mourning following the death of the bride's mother, Hippolita of Aragon. The child bride had arrived in Pavia late in 1488, but the news of her mother's death followed within a month, and the prospective bride and groom retired into the castle at Pavia to observe the customary period of mourning.

The earliest and most authentic description of the masque is to be found in an anonymous manuscript entitled "Feste in Milano nel 1490 (Paradiso)" which is preserved in the Biblioteca Estense of Modena.[94] Although the name of the author is not given, it is believed to have been the work of Giacomo Trotto, an *oratore* of Ferrara, who was present on the occasion and took a small part in it.

After describing the varied performances and dances which followed one another in breathless succession, the anonymous author went on to describe the masque which we quote in part:

After these dances were over, the music was stopped, at about twenty-four and one-half hours, and the performance began.
The "Paradiso" was made in the shape of a half egg,

whose inner side was made completely of gold, with a great number of lights resembling stars and with certain fissures in which were the seven planets, some high and some low, according to their degree. Around the upper edge of this half globe were the twelve signs [zodiac], with certain glass lights inside which made a gallant and beautiful sight: in this "Paradiso" there were many sweet, gentle songs and music.

Certain guns sounded and suddenly the silk cloth in front of the "Paradiso" fell away and a veil remained in front of it until a little child dressed as an angel had announced this performance. Without saying anything, the veil fell to the ground and there was such a majestic splendor and decoration that at first it seemed to be a natural paradise, and so it sounded, with the sweet music and songs which were inside. In the middle was Jupiter with the other planets nearby, according to their degree. After there had been singing and playing for a time, everything became silent: and Jupiter, with some good appropriate words, thanked the supreme God for having allowed him to create so beautiful, graceful, shapely and virtuous a woman for the world, as the most illustrious and excellent lady, the Duchess Isabella.

Apollo, who was below, marvelled at what Jupiter had said . . . and so he came down from Paradise with all the other planets and went to a mountain-top and from step to step these planets placed themselves nearby. As all were waiting, he sent Mercury to tell this Lady how he had come to earth to honor, exalt and extol her, to give her the three graces and to escort her with the seven virtues, . . . and so Mercury went to her Excellency and announced the arrival of Jupiter on earth with good words; and then he brought the reply to Jupiter. Having heard this and having understood why he had come to earth, all six planets thanked Jupiter one by one for the revelation that he had made so beautiful and comely a woman which he had created for the world, confirming in his desire to give her presents, and each of them, in order, offered a virtue. . . .

Some concept of the spectacle may be gained from the description provided by Bellincioni, who stated that the masque was named "Il Paradiso" because

. . . there was constructed there, with the great ingenuity and art of the Master Leonardo Vinci the Florentine, Paradise with all the seven planets that revolved, and the planets were represented by men, in shape and dress similar to those described by the poets, the said poets all speaking in praise of the aforesaid Duchess Isabella.[95]

The implication in Bellincioni's statement is that he wrote the verses to fit into Leonardo's scheme for the presentation. The rich mythological masque was enlivened by a multitude of sounds and verses by Bellincioni. The seven planets moved in their orbits amidst the diamond-powdered radiance of the hemispherical setting. As the action took place on stage, including the movement of the Olympian gods and the Graces and Virtues, stars flickered on and off. The sounds of Leonardo's mechanisms were said to be disguised by the soft voices of the ducal choirs.

Only one stage mechanism is illustrated and described in the surviving manuscripts of Leonardo. This is a drawing of a revolving stage which appears in the

[91] Leonardo da Vinci, *Manuscript B* [141], folio 58 recto. See also Ravaisson-Mollien *Manuscripts* [62].
[92] See footnote no. 43.
[93] Bellincioni, *Sonetti* [15], c. 148v. See also Bellincioni, *le Rime* [15], p. 208.
[94] *Feste in Milano* [145]. See also Solmi, "La Festa del Paradiso" [115], pp. 75–89 (Document XX).

[95] Bellincioni [16], *op. cit.* See n. 93 (Document XXI).

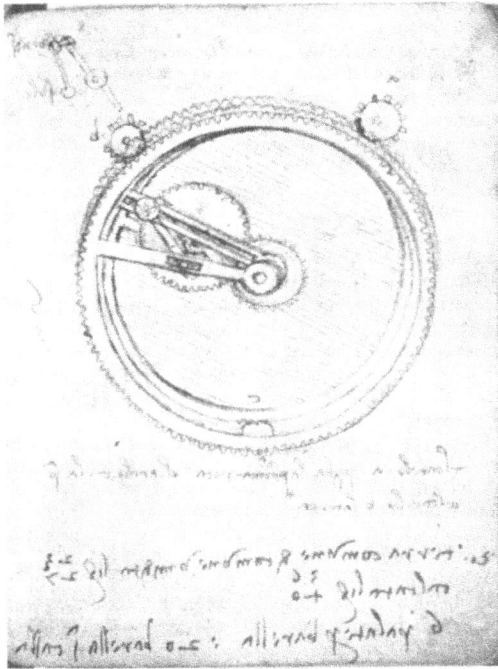

FIG. 28. Sketch by Leonardo da Vinci of the dial of Venus and the sun of an astronomical clock, probably the astrarium. Institut de France, MS L, folio 92ᵛ. Note Leonardo's mirror writing. Photograph courtesy of the Bibliothèque de l'Institut.

Arundel Manuscript [96] in the British Museum. It was probably designed by Leonardo for a presentation of Poliziano's "Orfeo," since the drawing and related notes date from 1506 to 1508, more than a decade after the production of "Il Paradiso" in Milan.[97]

The possibility that Leonardo was inspired by the astrarium is greatly strengthened by the discovery of sketches in his notebooks which may be of dials of the astrarium. The first of these was discovered

[96] Leonardo da Vinci, Arundel Manuscripts [142], folio 231 verso (Document XXII).

[97] Valletin, *Leonardo da Vinci* [75], pp. 201, 213 *et seq.* The diaries of the Venetian historian, Marin Sanuto, include a letter of a certain Galeazzo Visconti which was written from Amboise on the nineteenth of June 1518. The letter described the scenic representation produced at the castle of Cloux. The writer noted that the sun, moon, and planets were represented, the necessary apparatus being presumably similar to that constructed by da Vinci for the masque of "Il Paradiso" produced in 1490 at the castle of the Porta Giovio in Milan. Da Vinci is placed on the scene of the pageant within a week of its presentation by a note in his own hand in the *Codex Atlantico* where, in the midst of some geometrical calculations he noted that he was "on the twenty-fourth of June, St. John's Day, 1518, at Amboise in the palace of Cloux."

by Price and his findings were published in 1958. Price suggested that the sketch which appears on folio 92 verso of MS L of the Institut de France, may be a sketch of the astrarium's dial of Venus. This discovery is of considerable importance, because as Price has pointed out, this is the first of the many drawings of complicated mechanisms and machinery, which abound in Leonardo's papers, which may be identified with a device known to have actually existed.

The sketch depicts a mechanism having a series of three annular gears on its outer edge, each of which appears to have approximately 110 teeth. Mounted on an eccentric at the center is a geared linkage with a slotted bar mounted on an epicyclic gear wheel. Three pinions geared to the annular wheels are marked *a, b,* and *c*. The drawing is captioned "Venus, Sole [sun]." A fragmentary inscription which appears above the drawing has been translated: "The pinion *a* engages with the first second wheel, the pinion *c* with the first, and the pinion *b* with the third."

Price noted that the manuscripts of the de' Dondi astrarium do not include certain details which appear in this drawing, indicating that the sketch was almost certainly made directly from observation of the mechanism itself. For instance, the manuscript drawings illustrate but one of the three necessary pinions and annular gears, and the gear teeth behind the epicyclic circle are not shown. All of these items are clearly visible in Leonardo's sketch.

Although the identification of this sketch with the astrarium has been questioned by H. Alan Lloyd on the basis of the discrepancy in the number of teeth and the absence of the cranked arm for deriving the motion

FIG. 29. Dial of Venus of the model of the de' Dondi astrarium in the Museum of History and Technology, Smithsonian Institution, Washington. Photograph courtesy the Smithsonian Institution.

of the planetary image, Price believes that, inasmuch as the Leonardo sketch was obviously a quick sketch, it could not be expected to provide an exact representation. The only other dial of the astrarium depicting solar motion is that of the primum mobile.[98]

Subsequently one of the present writers (S. A. B.) noted a second drawing in the same manuscript resembling another dial of the astrarium. This sketch occurs on folio 93 verso, on the lower portion of the page immediately following the one on which the so-called Venus dial appeared. The second sketch most closely resembles the astrarium's dial of Mars. The drawing is very faintly outlined as if it might have been a preliminary sketch which was never completed, and it lacks the strong lines of the drawing of the Venus dial. In spite of its indefinite form, however, the resemblance to the Mars dial is unmistakable.

---

[98] Price, "Leonardo da Vinci and the Clock of Giovanni de' Dondi" [111] *Antiquarian Horology* 2, 7 (June 1958) : pp. 127–128; Lloyd, "Letter to the Editor," *Antiquarian Horology* 2, 10 (March, 1959) ; p. 199; Price, "Letter to the Editor," *Antiquarian Horology* 2, 11 (June, 1959) ; p. 222.

Leonardo da Vinci, *Manuscript L* [141] folio 92 verso and folio 93 verso.

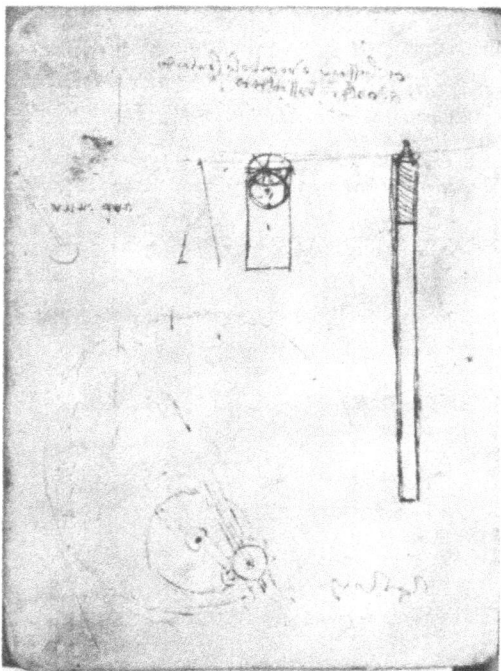

FIG. 30. Sketch by Leonardo da Vinci of a dial from an astronomical clock, probably the dial of Mars on the astrarium. Institut de France, MS L, folio 93ᵛ. Photograph courtesy the Bibliothèque de l'Institut.

Further information on these drawings is to be found in the writings of Professor Carlo Pedretti. In his work on *Leonardo da Vinci Fragments at Windsor Castle from the Codex Atlanticus,* Pedretti points out that these two sketches of clock dials from MS L can be compared to similar drawings which appear on folio 27 *v-a* and on folio 366 *r-a* and *v-b* of the *Codex Atlanticus.* The same drawings are repeated in folio 12,705 *r-v* of the fragments of Leonardo's manuscripts in the collection at Windsor Castle. This fragment, and the related drawings in the Codex, have been dated *ca.* 1498 by Pedretti and *ca.* 1497–1498 by Kenneth Clark.[99]

It is Pedretti's belief that the drawings relate not to the astrarium of de' Dondi, but to the famous astronomical clock produced by Lorenzo della Volpaia at Florence *ca.* 1480.[100] Although such a theory is tenable for lack of more definitive evidence, it seems perhaps more likely that the sketches are in fact of the astrarium's dials.

The discovery of this second sketch does much to confirm Price's tentative association of Leonardo with the de' Dondi masterpiece. Since the drawings are obviously not copied from a manuscript of de' Dondi's description of his astrarium, it must be assumed that Leonardo sketched the astronomical clock itself.

It may be possible at this time to arrive at a fairly accurate date for the sketches, assuming that they were not drawn from memory. If the assumption that the masque "Il Paradiso" was inspired by the de' Dondi

---

[99] Pedretti, "Nuovi documenti" [103a], pp. 57–58; "Il codice . . ." [103b], pp. 23–25.

[100] Pedretti, *Leonardo da Vinci Fragments* [57a], pp. 39–40; Clark, *A Catalogue* [24a], pp. 438–439; da Vinci, *Codex Atlanticus* [140], ff. 27v and 366r-a and 366v-b, *ca.* 1498. See note 134 below.

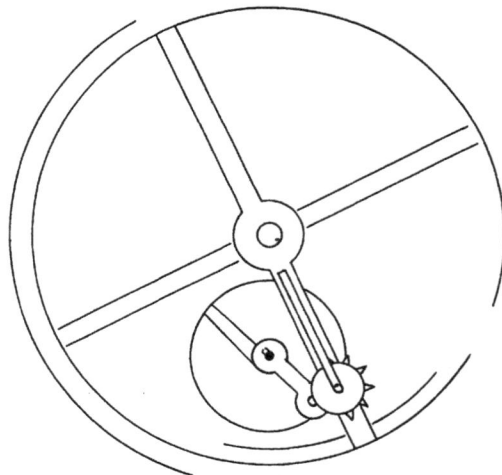

FIG. 31. Drawing from the sketch of the Mars (?) dial by Leonardo da Vinci reproduced in fig. 30.

FIG. 32. Dial of Mars on the model of the de' Dondi astrarium in the Museum of History and Technology, Smithsonian Institution, Washington. Photograph courtesy the Smithsonian Institution.

mechanism is acceptable, the sketches of the two dials may have been made as early as the end of 1489 or the beginning of 1490, the period during which Leonardo was visiting the library at the castle in Pavia so frequently.

On the other hand, the sketches may be as late as— although no later than—1494 to 1496, the years when Leonardo was engaged at Vigevano for Duke Lodovico il Moro.[101]

In about 1492 Duke Lodovico had undertaken to rebuild the little city of Vigevano, where he had been born and where he continued to spend his summers. Part of his ambitious plan was to reconstruct the gothic Castello Visconteo, built in 1340, which dominated the city. He employed Leonardo da Vinci to provide plans for the castle's enlargement and reconstruction.[102] At the same time the Duke appointed Donato Bramante, engineer of the city of Milan, to work with Leonardo in the reconstruction. The Duke's plan for modernization of the city included widening of the town square by the removal of a number of the old houses. He instructed Bramante to design and execute an arcaded piazza in the center of the city. New buildings were constructed and ornamented with fresco paintings and the elegant square—decorated in many colors by Bramante himself to resemble a grand hall with a loggia—survives to the present, although much restored.

The old Castello Visconteo was rebuilt at the same time that the square was being constructed. The im-

petus for the modernization of the castle resulted from the bitter complaints voiced by Duchessa Isabella of Aragon concerning the discomforts she had suffered from the inconveniences of the castle in the first years of her marriage.[103]

Leonardo's reconstruction included ambitious plans for redecoration of the great hall. He proposed to paint twenty-four scenes from Roman history, together with portrait busts of the philosophers as well as other motifs, between pilasters of blue and gold. He submitted an estimate of cost of fourteen lire for each Roman scene and ten lire for each philosopher, although the paint alone would probably cost him seven lire. Despite the modest bid, the project was apparently abandoned. Leonardo did, however, decorate a series of small rooms or *camerini* in the Duchess's quarters which had been modernized. During the period that the extensive renovation of the castle and city was in progress, the Ducal resources diminished and his artists were not paid. This situation led Leonardo to leave the Duke's service in June 1496. Unable to find other employment, however, he was forced to return and continue his decoration of the palace.[104]

It was during the period while the reconstruction at Vigevano was in progress that a startling note was recorded in the history of the astrarium. In a letter addressed to Gualtiero Bescapé dated 6 November 1494, Duke Lodovico il Moro informed him:

Having learned that which you have brought to my attention by means of Giovanni Giacomo Gilino, of the Astrolabio [astrarium] which is in the hall of Rosate, I tell you to let it remain there until we are in Milan, because I will then order what we wish to have done about it.[105]

This is the first documentary evidence that the de' Dondi masterpiece had ever been removed from the ducal library in the castle at Pavia. It is apparent also from the Duke's letter that its whereabouts had not been

---

[101] Reti, "Non si volta . . ." [112a], pp. 26–28.
[102] Uzielli, *Ricerche* [73a], p. 612.

[103] Malaguzzi Valeri, *Bramante e Leonardo da Vinci* [50].
[104] Leonardo da Vinci, *Atlanticus* [140], folio 380 recto; Vallentin [75], *op. cit.*, pp. 212–214.
[105] *Carteggio Generale* [128] (Document XXIII).

FIG. 33. View of the Castello Sforzesca at Vigevano, showing the Falconry. From a photograph by the author, S.A.B.

known to him until the time of his writing. Some clarification may be derived from the fact that Duke Gian Galeazzo Sforza, Lodovico's nephew with whom he had jointly ruled, had died in October 1494. Gian Galeazzo, who lived in the Castello Visconteo in Pavia, was the owner of the astrarium during his lifetime. Lodovico may have inherited it after his nephew's death, or otherwise desired to acquire it. The fact that his letter regarding the astrarium's recovery was written not more than a week or two after Gian Galeazzo's death seems to verify this conclusion.

In this connection the passage quoted above (p. 26) from the work of Malaguzzi Valeri,[106] in which he noted the clockmakers who attempted to repair the astrarium, is quite important. In the last sentence quoted he stated that the mechanism was then dismantled and removed from the library of the castle at Pavia and shipped to a hall in the castle of Rosate.

The reference to the castle at Rosate is of considerable significance. The castle was in the fief of Rosate, a small community situated southwest of Milan adjacent to the boundary of Pavia. This "feudo," or property, had been awarded by Duke Lodovico il Moro to Ambrogio Varese, his court astrologer and personal physician,[107] on 11 November 1493.

The date of transfer of the astrarium to Rosate cannot be determined with any degree of accuracy although presumably it was after the transfer of the property to Ambrogio. Some clue may exist in the files of Ambrogio Varese's papers in Milan,[108] which remain to be examined.

Amplification of the mystery results from subsequent correspondence relating to Bramante's decoration of the castle at Vigevano.

The one part of the castle, in addition to the Piazza Ducale, which Bramante is known with certainty to have designed and constructed was the Palazzo delle Dame or "The Ladies' Palace" attached to a wing of the castle. In addition to the design and supervision of the construction itself, Bramante personally painted colorful frescoes for the Duchess on the walls and ceilings. Some concept of the extent of Bramante's fresco decorations for the Palazzo may be derived from a surviving letter dated 4 March 1495, addressed to the Duke from Bianchino da Palude at Vigevano. Bianchino reported on the progress of the work and stated that Bramante had gone to Pavia to obtain several items from the son of Master Ambrosio da Rosa in order to be able to work in the room of the round ceiling. Meanwhile, the wooden eaves of the room had been furnished and covered. The report continued with a description of the work in other rooms:

FIG. 34. View of Castello Marco Visconti at Rosate, remnant of a tenth-century castle once owned by Ambrogio Varese (1437–1515), court astrologer.

My most illustrious, excellent and only Lord.—[That] Your Excellency may know how the decoration of the castle is progressing. The repair of the gesso is finished in the room which Bramante is painting, near the street. The room near the chapel will be painted. The room with the domed ceiling is not ready to be worked on yet. Bramante went to Pavia to get some things from the son of Maestro Ambrosio da Rosa so that he could work on that room. The wooden eaves of these rooms are furnished and covered. The bedroom and wardrobe [dressing room?] of the most illustrious Duchess will be furnished this week. The hall below the balcony is plastered and the floor of stones has been begun. The room in this hall is plastered and the floor has been laid. . . . I continue to recommend myself to Your Most illustrious Lordship.

Vigevano, 4 March 1495.

Your most faithful servant,

Bianchino da Palude.[109]

This communication should be considered with another surviving letter addressed to the Duke by the custodian of the ducal castle in Pavia, Jacopo da Pusterla. It is dated from Pavia on the following day, 5 March 1495, and stated:

My most Illustrious and Excellent Master:
There has been here Bramante, engineer of Your Excellency, who states that he has been commissioned to obtain

---

[106] Malaguzzi Valeri [49]. See note 63.
[107] Negri, *Rosate* [57], and Gabotto, *Nuove ricerche* [35]. See *Varese, Ambrogio* in Appendix II, *Biographical Index.*
[108] *Feudi, Rosate* [129] 1, c. 182 v. See also *Autografi, Medici* (Ambrogio da Rosate) [130].

[109] *Carteggio degli Ambasciatori Estensi* [134], Env. 11ᵃ. Letter from Bianchino da Palude to Duke dated 4 March 1495 (Document XXIV).

various sketches from the clock which is in this library, of various planets to decorate a certain ceiling of a room at Vigevano; and I, not having had other instructions from Your Excellency, did not permit him to remove out of the aforesaid library any designs until he had special permission to do so. I hope that you will care to advise me as to what I am to do by means of a letter signed in your own hand. In anticipation, Your Excellency, I continue to recommended myself. From Pavia 5 March 1495.

E. EX. V.
Your most faithful servant,
Jacopo de Pusterla,
Governor of the castle etc.

(To Duke Lodovico Maria Sforza etc.).[110]

[110] *Registri Ducale* No. 121, Autografi C. 98—Bramante [131] (Document XXV). This letter regarding Bramante's visit to Pavia to study the de' Dondi astrarium has been noted and quoted in the following works:

a. Cicognara [24] 3: pp. 294-296.
b. Vedova, *Biografia* [76] 1: pp. 339–344.
c. Piertrucci, *Biografia* [59], p. 106.
d. Clausse, *Les Sforza* [25], p. 367.
e. Barni, *Vigesimum* [4], pp. 161–162.
f. Magenta, *I Visconti* [48], p. 570.

There also survives the transcript of the Duke's reply from Milan dated the following day, 6 March 1495, which commented as follows:

Custodian of the Castle at Pavia
We are happy to have you permit to be taken by our engineer Bramante those sketches of the clock in that library which he likes and so with this we give you complete permission for it.

Milan, 6 March 1495.[111]

This correspondence is further supplemented by another letter noted by M. Caffi [112] in which the Duke Lodovico il Moro ordered his "engineer Bramante" to betake himself to Pavia to obtain some designs of "the clock which is in that library" because they would be useful to him in reproducing the representations of the planets on the ceiling of a room in the castle of Vigevano.

[111] Missive, *Registro Ducale* [132], Libro 193, folio 211 **verso**, dated 6 March 1495 (Document XXVI).
[112] Caffi [85], p. 550. Missive [133], Libro 37, folio 90.

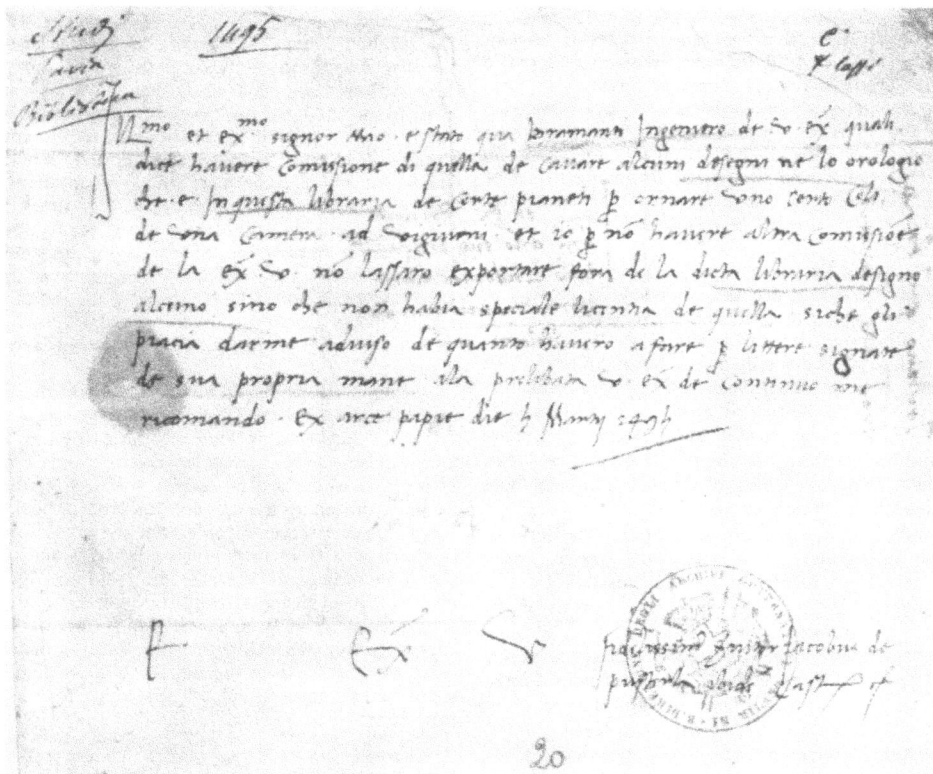

Fig. 35. Letter from Jacopo da Pusterla to Duke Lodovico dated 5 March 1495, informing him of Bramante's visit to the ducal library, where he sketched the astrarium. Archivio di Stato, Milan. *Registro Ducal N. 121, Autografi C. 98.* Photograph courtesy Archivio di Stato, Milan.

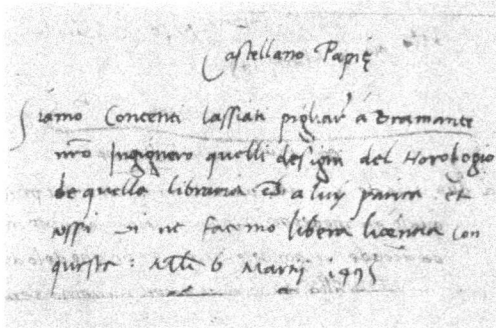

Fig. 36. Autograph note dated 6 March 1495, from Duke Lodovico to Jacopo da Pusterla granting permission to Bramante to make sketches of astrarium. Transcript, *Missive, Registro Ducale Libro 193, folio 211*. Archivio di Stato, Milan. Photograph courtesy Archivio di Stato, Milan.

From this correspondence relating to the astrarium, it is readily apparent that between the dates of November 1494 and March 1495 the astrarium was restored to its usual place in the ducal library at Pavia.

Some interest is to be found in a phrase which occurs in the letter of 4 March 1495, from Bianchino da Palude to the Duke, quoted in the foregoing, in which the writer noted that ". . . Bramante has gone to Pavia to get some things from the son of Master Ambrosio da Rosa so that he could work on that room. . . ."

He referred to the room with the domed ceiling, presumably the one in which the planetary fresco was to be painted. Yet, the letter of the custodian of the Castello Visconteo dated the following day described the astrarium as being still in the ducal library. The astrarium at this date is conclusively established as being in the ducal library at Pavia, and no longer at Rosate. Although it is not known what Bramante may have needed from the son of the court astrologer, it may have been astrological or astronomical references to supplement the design for his fresco.

Although it is not possible to determine the form of Bramante's planned or accomplished fresco based on the de' Dondi astrarium, several possibilities may be considered. He may have combined designs of the courses of the planets derived from the seven dials of the mechanism in some decorative manner. It is far more likely that he represented Duke Lodovico Sforza and his family as patrons of learning with the astrarium carefully delineated and prominently displayed.

These letters cannot but arouse great curiosity as to whether the fresco produced by Bramante has survived. A careful examination of the published and manuscript works relating to the castle at Vigevano yielded most

discouraging results. There is substantial evidence that the frescoes and other decorations of the castle were all destroyed, probably as early as the sixteenth century. Although the *logetta* or colonnaded gallery survives, the Palazzo delle Dame perished in its entirety. It is believed that this wing was destroyed when a large part of the castle had to be rebuilt in about 1548. The cause is attributed to extensive damage resulting from Bramante's hasty and poor construction. According to Casati,[113] Bramante employed mattings of canes covered with stucco or plaster made of chalk and lime for the construction. Simon dal Pozzo blamed Bramante for the damage because

This master architect [Bramante] had the vice that his structures had habitually poor foundations, as in effect may be seen in that part of the palace [Palazzo delle Dame] in the garden towards the covered roadway where there are the most beautiful rooms which displayed many fissures. If Ferdinando Gonzaga, governor of the State for Emperor Charles V, had not had new foundations made for them in 1548, they would have fallen, and this is true also of many places in Rome. . . .[114]

Probably the greatest degree of damage to the castle has resulted from its occupancy as a *caserma* or military barracks for more than a hundred and thirty years. The castle has been constantly maintained as a military post from 1829 to the present. According to Barucci,[115] the constant occupation of the premises by soldiers over such an extended period explains the vandalism and the countless vulgar adaptations which have changed the castle in the course of time.

If any of the decorations by Bramante had survived the earlier reconstruction, they were probably ruined or removed during the remodeling accomplished between 1854 and 1857 under the direction of the military engineer, Colonel Ludovico Inverardi, who was generally believed to have exercised bad judgment and worse taste.

Although the public is not permitted in the military establishment, in 1963 one of the present writers (S.A.B.) succeeded in entering the castle and making a comprehensive tour, room by room. Some fragments of frescoes which can be seen in the arch of the archducal entrance into the city are of a later period. Other fragments of frescoes of an earlier period, and possibly of the time of Bramante, survived at least until 1913 as noted by Malaguzzi Valeri,[116] but they are no longer visible. No evidence of the fresco decorations of Bramante remains. If Bramante did, in fact, create a fresco decoration for the ceiling of the Palazzo delle Dame based on a motif from the de' Dondi astrarium, and it seems certain that he did, it must have perished more

[113] Casati, *I Capi di' Arte* [23]. See also his Resoconto [86].
[114] Barucci, *Castello* [8], pp. 66–71. Cites Simon Dal Pozzo, *Libro grosso dell'Estimo generale*, MS. compiled 1548, folio 535.
[115] Barucci [8], pp. 66–67. See also Malaguzzi Valeri, *Bramante* [50], pp. 167–168.
[116] Malaguzzi Valeri, *Bramante* [50], op. cit., p. 657.

than a century ago with the destruction of that part of the structure.

The conditions in the castle reported by Malaguzzi Valeri remain to the present. The once elegant halls are lined with camp beds and gun racks. The former long vistas of arches and loggias are today walled up with brick, and here and there unsightly windows and galleries have been added. Lines of barracks and shelters for horses and military equipment are attached haphazardly to the original ancient edifice.

A search of Italian archives for cartoons or sketches which Bramante may have made in preparation for the application of the fresco has been unsuccessful. Such sketches may possibly survive in some Italian library and come to light at a future date. Until then, the only available contemporary illustrations of the de' Dondi astrarium are Leonardo da Vinci's sketches and the drawings which form part of the various manuscript copies of the de' Dondi *Tractatus astrarii* or *Opus planetarium*.[117]

The eventual fate of the de' Dondi astrarium remains a mystery, although it survived at least partially until 1529 or 1530. The correspondence between Jacobo da Pusterla and Duke Lodovico relating to Bramante's examination of the astronomical clock are sufficient proof that the mechanism remained in the Visconti castle at Pavia as late as 1495, and that until that time it was on display and not stored away.

THE ASTRARIUM IN THE SIXTEENTH CENTURY

The last historical records we have of the de' Dondi astrarium date from the sixteenth century. Subsequently, it was rather de' Dondi's fame as a man of learning which was remembered. Bernadino Scardeone, who devoted a chapter of his *De antiquitate urbis Patavii & claris civibus Patavis . . .*, (Basle, 1560),[118] to Giovanni de' Dondi begins his account thus:

Another outstanding ornament of his country, and of his age, was Joannes Dondus, an illustrious philosopher, a physician, and a most excellent mathematician. He was a Paduan, by descent and origin, like his father, but it befell that he was born at Chioggia, because his father, Jacobus, happened then to be working in that town as a doctor. There survives a very famous work of his, divided into three volumes and adorned with marvelously skillful lines and diagrams, concerning the construction of a planetarium, wherein the course of the celestial bodies is easily measured by suspended weights, as is that of the hours, just as his father had previously done in his clock. I saw the first part at the house of Galeatius Horologius, a most distinguished citizen. The title of this work is *The Planetarium of Joannes Dondus, a Citizen of Padua*. Of this work there are three volumes, in the first of which are instructions for putting together the planetarium, and there

are 25 chapters. I hear that the remaining two volumes are with other pupils [? or members] of the same family. This Joannes was a great friend of Franciscus Petrarcha. . . .[119]

The manuscript mentioned by Scardeone is probably one of those listed in de' Dondi's widow's inventory of her husband's property (see above, p. 19). It might be the manuscript which remained in the family until 1795 (see below, p. 41, no. 1), but this manuscript refers to de' Dondi's clockwork as *astrarium*, not *planetarium*. "Galeatius Horologius" could reasonably be identified with Giangaleazzo Dondi dall' Orologio, a son of Giovanni by his second wife, Catterina, though Giangaleazzo would hardly have been still alive when Scardeone saw the manuscript at his house. Scardeone's reference to "three volumes" of de' Dondi's work is clear enough; the work consists of three parts, describing, respectively, the construction of the astrarium, its operation and adjustment, and the correction of errors in its movement, and at this time the second and third parts must have been separate fascicules. What is most curious, is that Scardeone, a Paduan born in 1478, and familiar as he seems to have been with de' Dondi's life, written work and family, should say nothing of the astrarium itself.

The astrarium appears to have survived (though not in working order, or even in a state of completeness) until the end of the third decade of the sixteenth century. It seems, however, that the fact of its construction by de' Dondi was beginning to be forgotten. Three sixteenth-century works give accounts of Charles V's interest in an astronomical clock (almost certainly the astrarium) in 1529 or 1530 while he was in Italy for his coronation at Bologna as King of Lombardy. One of these accounts attributes the astrarium to another inventor, another says implicitly that its author is unknown, and the third says nothing of its constructor.

In 1550, some eight years before the publication of Scardeone's biographical note on de' Dondi, Girolamo Cardano wrote in his *De subtilitate . . .*:

Ianellus Turrianus of Cremona, whom we have mentioned above, a man of sharp intellect, both invented many such things and improved such things invented by others. When I brought to light (for, by good fortune, I was born to restore not only those worthy crafts to which I apply myself, but also those I encounter casually) a certain machine representing the whole world, made by Gulielmus Zelandinus and since fallen to pieces and decaying in darkness, Ianellus completely restored it. Taking it as a model, he made, for the Emperor Charles the Fifth another machine such that you can see both the moments of time and the individual parts of the signs, and also the very slow motion of the eighth sphere. It is possible to perceive in it those various divisions of the circle of the signs [*sc.* zodiac] which are called houses [i.e., the 12 astrological houses,], the equal and unequal hours, and what is more wonderful, it serves for all parts of the world, so that the machine really depicts the whole universe. I omit the progression and regression

---

[117] Surviving manuscripts of the *Tractatus astrarii* and of the *Opus planetarium* are listed and described in the section entitled *Manuscripts of the Astrarium*, pp. 40–43.

[118] Scardeonius, *De antiquitate urbis Patavii* [67], pp. 206–207 (Lib. II, Classis IX), "De Ioanne Dondo seu Horologio."

[119] (Document XXVII). For de' Dondi's friendship with Petrarch, see above, p. 15.

FIG. 37. Portrait of Charles V, Holy Roman Emperor from 1519 to 1556. From a sixteenth-century mezzo ducato of Naples in the numismatics collection of the Museum of History and Technology, Smithsonian Institution, Washington. Photograph courtesy Smithsonian Institution.

of the wandering stars [i.e., the planets], the latitudes and the altitudes, and innumerable other things, so that, in short, it is not something less when described than it is believed to be.[120]

[120] Hieronymus Cardanus (Girolamo Cardano), *De subtilitate* . . . (Lyons, 1554) [22], pp. 611–612, Lib. XVII "De artibus," under the marginal catch-phrases, "Ianellus Turrianus Cremonensis" and "Gulielmus Zelandinus autor sphaera coelestis mirabilis artificij." This is a revised edition of the various books of the *De subtilitate* . . . , ("[libri] nunc demum ab ipso autore recogniti atque perfecti"), Cardano's preface to which is dated at Paris, "in itinere," 1552. The passage follows a description of clocks without ropes (i.e., spring-driven clocks), and a seat, made for the Emperor, employing a gimbal mounting (the so-called Cardan suspension). As well as some subsequent material, it is not found in the first editions of the *De subtilitate* . . . , Paris, 1550, and Nuremburg, 1550, nor in the Paris reprint of 1551. In these editions the passages on clocks and the seat conclude with the following sentence, "Ianellus Turrianus Cremonensis, cuius etiam supra meminimus, horum inventor est" (Paris, 1550 and 1551), p. 268, a statement of some interest, perhaps, to historians of the Cardan suspension, for Cardano also refers to the use of the suspension for oil lamps as if this were generally known. We have not attempted a comparative study of the many editions (some pirated; see Oystein Ore, *Cardano. The Gambling Scholar,* Princeton, N. J., 1953, p. 43) of the *De subtilitate* . . . , but can say that the passage inserted in the 1554 edition is also found in the French translation (Paris, 1556, pp. 322–323), the Latin edition (Lyons, 1559, pp. 579–580), and the *folio* Latin edition (Basle, 1560, p. 453). In the *octavo* Latin editions (Basle, 1560, p. 1029; 1582, p. 1148; and 1611, p. 1148), and also in the edition (? Basle, 1582) reprinted in the ten-volume *Opera omnia* of Cardano (Lyons, 1663: 3: p. 612), the same passage occurs, but with the reference to Torriano omitted, beginning simply, "Nuper etiam quidã machinam illam mundi uniuersalem olim à Gulielmo

It seems reasonable to assume, especially if this passage be read in conjunction with the translation from Sacco below, that Cardano is referring to the de' Dondi astrarium. His attribution of it to Gulielmus Zelandinus suggests that this mysterious clockmaker's name had become traditionally associated with the astrarium. Hence it confirms, perhaps, that Gulielmus did repair the astrarium in response to the appeal from the secretary of Duke Fillippo Maria Visconti in 1456 (see above, p. 25).

Cardano's statements that it was he who brought the forgotten machine to light and that it was, in fact, repaired by Torriano are not substantiated by an account published fifteen years later by Bernardo Sacco in his *De Italicarum rerum varietate et elegantia* . . . :

. . . During the reign over the regions beyond Padua [i.e., Lombardy] of Joannes Galeacius Vicecomes [i.e., Visconti], it is reported that there was completed a clock, showing not only the hours, but also the stars [i.e., the planets], and the courses of the Sun and the Moon. The author of this work is not known. The clock was set up in the stronghold, or castello, of Pavia, where, after the death of the Prince, a work as marvellous as this lay despised, and even some of its circles [? dials] were removed. In the very next century, in the 1500's, about the 29th year, when Charles the Fifth received the imperial crown at Bologna, that clock, incomplete (as it was) and marred with rust [*or* corrosion], was brought from its place to the aforesaid Emperor, who, having inspected and admired the machine, ordered the repair of so great a work

Zelandino fabricatam atq; dissolutam. . . ." The text translated by Montañes [93], pp. 8–9, from the Antwerp, 1611, edition differs substantially from either version, but we have not seen this edition.

In Lib. I of the *De subtilitate* . . . , at the beginning of a description of the "Machina Ctesibica," Cardano wrote, "Hic igitur demonstratis tãquã principiis, ratio cõsurgit machinę Ctesibice, quę sic cõstat vt etiã Ianellus Turrianus Cremonensis vir magni ingenij in omnibus que ad machinas pertinent, opere ipso expressit" (Lyons, 1554 ed., p. 11). This passage is also found in the editions of 1550 (Paris, p. 6), 1551 (p. 6), 1556 (p. 6), 1559 (p. 18) and the folio Basle ed. of 1560 (p. 9) but is abbreviated, to stop at ". . . sic constat," in the octavo Basle ed. of the collected works (3: p. 361), thereby omitting any reference to Torriano.

We are indebted to Dr. C. H. Josten for drawing our attention to the following interesting passage in the "mathematicall Praeface" which John Dee (1527–1608) completed at his "poore House At Mortlake" on 9 February, 1570, for the first English translation of Euclid (H. Billingsley, trans., *The Elements of Geometrie of the most auncient Philosopher Euclide of Megara* [London, 1570], sig.ciiij*): "By Wheles, straunge workes and incredible, are done: as will, in other Artes hereafter, appeare. A wonderfull example of farther possibilitie, and present commodities, was sene in my time, in a certain Instrument: which by the Inventer and Artificer (before) was solde for XX. Talentes of Golde: and then had (by misfortune) receaued some iniurie and hurt: And one *Ianellus* of *Cremona* did mend the same, and presented it vnto the Emperor *Charles* the fifth. *Hieronymus Cardanus,* can be my witnesse, that therein, was one Whele, which moued, and that, in such rate, that, in 7000. yeares onely, his owne periode should be finished. A thing almost incredible: But how farre, I keepe me within my bounds: very many men (yet aliue) can tell." It is curious that Dee makes statements which are not found in Cardano's printed text.

by workmen called from all parts. While they worked in vain on the task of reconstruction, there arrived a certain Joannes of Cremona, called Ianellus, of misshapen appearance, but bright of mind, who when he examined the work said he could repair the machine, but this was of no use to anyone as the iron parts [121] were brittle with rust [or, corrosion] and eroded away, unless a new instrument, similar to the old one and of the same proportions, be put together. Having set about this work, he completed the task by daily labour, imitating, emulating and equalling the earlier device. When the Emperor wanted to take it to Spain, he took Master Ianellus at the same time. . . .[122]

A third account of this episode was given by Stefano Breventano in his *Istoria della antichità, nobilità, et delle cose notabili della città di Pavia,* published at Pavia in 1570, when he described the state of the castle at Pavia:

This palace has four large towers but of them there exist today only the two on the side toward the city; the two which overlooked the park were (as we have said) demolished by Lautrec the Gascon with artillery, and on that which on entering the said castle was on the right there was in my time a clock of marvellous construction, which had been made to the order of Duke Giovanni Galeazzo Visconti, which not only with a sign [= index, *or, perhaps,* dial] and the sound of a bell showed the hours, but even gave all the courses and revolutions of the planets and celestial signs. This [clock], not having been cared for as a result of changes in the state, corroded with rust, and with its wheels removed from their [proper] places, fell completely into ruin, and [the wheels] thereafter were collected [together] by a Master Gianello of Cremona, a man of the greatest ingenuity in that art, [who] at the request of the Emperor Charles the Fifth, made another that resembled it.[123]

Parts of this passage may well be paraphrases of the accounts of Cardano and Sacco. It is quoted here, however, because Breventano says that the clock survived in his time and, if we may infer from this that he remembered it, then possibly he was not in error when he said that it struck the hours, though he was wrong in saying it was *on* the tower. No other evidence of this striking mechanism is forthcoming, but it is not incompatible with de' Dondi's "common clock" mechanism (see above, p. 11) which de' Dondi did not describe in his treatise on the grounds that anyone capable of understanding the rest of the astrarium would certainly know how to construct this part.

Some conception of the appearance of the Castello Visconteo at Pavia, when Charles V was in Italy, may be derived from the account in 1636 of Giovanni Battista Pietragrassa who, perhaps following Breventano, reported that as of the year 1527:

. . . a large part of the walls had been levelled (by the artillery of Lautrec) as well as the two towers of the castle which overlooked the park, in one of which had been maintained the library. . . . In another of the towers noted was placed a clock, which was, because of its marvellous design and inestimable construction, most cunning, the fragments of which had been dispersed for a long epoch, and consumed by corrosion, Tonello [*sic*] of Cremona, architect of Emperor Charles V and highly reputed geometrician, collected and imitated it well, adding also more than the original had. In its art and industry he surpassed even the original with the one he was making; among these it was reputed to be almost a microcosm, a superhuman work, and it was preserved in Toledo by Filippo, Catholic king of Spain and second son of the Emperor. It was held in great esteem, and in it the stars of the heavens were so smoothly discovered that it seemed superhuman rather than otherwise. . . .[124]

No reference to the astrarium of de' Dondi has been found subsequent to 1529 or 1530, and attempts by modern scholars to obtain some statement of its disposal have proven vain.

According to an often repeated story (which Basserman-Jordan,[125] Baillie,[126] and Lloyd[127] have accepted), the de' Dondi astrarium was presented to the Emperor Charles V, was repaired by Gianello Torriano (see below, p. 57), and was taken to Spain by the Emperor, accompanied by Torriano. The astrarium was then at the Convent of San Yuste in Estremadura whither the Emperor retired in 1557 after his abdication and it was still there in 1809 when, during the Peninsular Wars, Marshal Soult ordered the burning of the Convent, with the consequent destruction of the astrarium. The accounts of Cardano, Sacco, Breventano, and Pietragrassa, are not contradicted by any early source, and suggest strongly that the astrarium remained in Italy (as, indeed, Morpurgo[128] has maintained). The currency of the story is the more surprising as the sixteenth-century accounts were discussed at length in a nineteenth-century study by Escosura[129] and a recent article by Montañes.[130]

Whether the astrarium perished completely has never been satisfactorily determined. It is clear from the description here quoted that the astrarium was in a very bad state of preservation when seen by Charles V, and it is indeed probable that Torriano, and other clockmakers, who examined it, dismantled it further in

---

[121] The mention of iron parts and, perhaps also, of rust conflicts with Philippe de Maisieres' statement [90], that de' Dondi made the astrarium entirely of brass and copper (see above, p. 20 and Document V).

[122] Saccus, *De Italicarum rerum varietate* [64], pp. 150–151 (Lib. VII, Cap. XVII), "De horologijs nostrae tempestatis, antiquioribus ignotis"—p. 150 (Document XXIX). The first edition of this work was published in Pavia in 1565.

[123] Breventano [19], Libro I, folio 7ʳ (Document XXX).

[124] (Pietragrassa), *Annotazioni diverse* [60] (Document XXXI).

[125] Basserman-Jordan, *Die Geschichte der Räderuhr* [12], p. 45, citing Erizzo [34] (see below, p. 45 and n. 161).

[126] Baillie, "Giovanni de' Dondi . . ." [81]. See n. 2.

[127] Lloyd, "Horological Masterpiece" [97], p. 71; *idem., Some Outstanding Clocks* [43], p. 24.

[128] Morpurgo, *Dizionario* [55], p. 61; *idem.* in Barzon, Morpurgo, Petrucci and Francescato, *Tractatus astrarii* [11], p. 41.

[129] Escosura, "El Artificio" [91], *passim.* Escosura's article is an important but neglected study of the life and work of Gianello (Juanelo in Spanish) Torriano, especially of his great hydraulic works.

[130] Montañés, "Los Relojes del Emperador" [93].

their attempts to understand its functioning, and perhaps even removed some components. The occupation of Italian palaces and convents by military personnel—a practice initiated early in the nineteenth century and persisting to the present time—has done much to dispossess and destroy their treasures as well as their discards. Yet, whatever remained of the astrarium would not easily be overlooked in a dust heap. Some remnants that survived may even have been put to use in some other manner. It is, perhaps, conceivable that, from the confusion and cluttered storage of some Italian castle there may someday be removed a recognizable fragment of the original astrarium of Giovanni de' Dondi.[131]

[131] An example of the reuse of part of an astronomical clock is, perhaps, the late gothic astrolabe, ?Italian, ?c. 1400, diam. 162 mm., in the Museum of the History of Science, Oxford, no. 57-84/175 A. This astrolabe, clearly a composite instrument, has a *rete,* of which the circumference (i.e., the band which would normally represent the Tropic of Capricorn but which in this case does not touch the southern point of the ecliptic circle) is cut with 120 teeth and divided equally into twelve parts named with the months of the year. The teeth could have served no purpose on the astrolabe in its present form, and it seems probable that this *rete* formed part of an astronomical clock, made not long after de' Dondi's astrarium; see F. R. Maddison, *A Supplement* [47], pp. 32–33, and pls. XXIV and XXIIb.

Fig. 38. *Rete* of a late Gothic astrolabe, perhaps Italian, *ca.* 1400, in the Museum of the History of Science, Oxford, No. 57-84/175A. The *rete* may have formed part of an astronomical clock made shortly after the astrarium of Giovanni de' Dondi. Photograph Edmark, Oxford.

The astronomical clock, inspired by de' Dondi's astrarium, and made for the Emperor Charles V [132] is, perhaps that described by Ambrosio de Morales in a book published in 1575,[133] though this description may refer to a later and more elaborate instrument. Morales says that Torriano told him he had spent twenty years designing the clock and three and a half years making it by hand; "the clock had all of 1800 wheels, without [counting] many other things of iron and brass that are involved." If this is so, then this clock must have been completed in Spain. Charles V's horological collections, did not remain at the Convent of San Yuste after the Emperor's death. The clock under discussion appears to have been seen at Toledo, in Torriano's lifetime, by Esteban de Garibay (*cf.* Pietragrassa's remark above) and is listed in an inventory compiled in 1602 after the death of Philip II, whose service Torriano had entered following the death of Charles V.[134]

MANUSCRIPTS OF THE ASTRARIUM

At least eleven manuscripts, all written in Latin, of Giovanni de' Dondi's description of the construction and operation of his astrarium are now known. These manuscripts have been described singly and severally by various writers, but never all at the same time with individual descriptions.

As mentioned above (p. 15), the terms "astrarium" and "planetarium" were both used to describe de' Dondi's masterpiece. Of the eleven known manuscripts, the titles of two, the text of another, and the colophons

[132] The date when Torriano entered the service of Charles V has been queried by Morpurgo, *Dizionario* [55], p. 186.
[133] Morales, *Las Antigüedades* [54], pp. 91–93; quoted by Escosura [91], *op. cit.,* pp. 20–21, and Montañés [93], *op. cit.,* pp. 6–7 (*cf.* also the brief description of the clock quoted by Montañés from a life of Charles V, written by Snouckaert (Zenocarus) and published in 1559); a partial English translation of Morales' account is found in Woodbury, *Gear-Cutting Machine* [79], pp. 45–46.
[134] Morales, in his account of Torriano's clock mentions an astronomical clock by "another Italian of these times" which was mentioned in "a letter from Hermolao Barbaro to Angelo Politiano." This latter statement appears to be inaccurate and the reference is almost certainly to a letter from Politian to Francesco della Casa, dated vi Id. Aug., 1484. Lest it be assumed that the clock Politian mentioned was the astrarium, the beginning of his letter will be quoted here: ". . . accepi epistolam tuā, qua mihi significas, allatū istuc esse de machinula Automato, quae sit nup à Laurētio quodā Florentino constructa, in qua siderum cursus, cum coeli ratione congruens, explicetur. . . ." (*Omnia opera Angeli Politiani, et alia quaedam lectu digna* [Venice, 1498], sig. f1ʳ-1ᵛ). The clockmaker referred to is Laurentius Vulpariae (Lorenzo della Volpaia, b. 1446; see Morpurgo, *Dizionario* [55], pp. 202–204) and the clock was made for Lorenzo de' Medici (for a translation of Politian's fairly detailed description of the clock see Baillie [5], pp. 9–10). It might here be mentioned that both della Volpaia's clock and the astrarium of de' Dondi, are mentioned (and useful references given) in Beckmann, *A History of Inventions* [13] (the 4th English edition, and the best). The section of this work on clocks and watches (1: pp. 340–373) and on waterclocks (1: pp. 82–86), though confused, contain many neglected references which should be systematically reviewed.

of two use the word "astrarium," and four have the word "planetarium" in the title.

Three of the manuscripts are believed to be four-teenth-century copies, and eight to be fifteenth-century copies.

Scholars appear to be in agreement that the earliest of the manuscripts are MS. D. 39 of the Biblioteca Capitolare Vescovile in Padua, and Cod. 85 of the Biblio-teca Nazionale Marciana in Venice. The only one of the eleven manuscripts to have been fully pub-lished and edited is that in the Biblioteca Capitolare Vescovile in Padua (see above, p. 14), so these views must be regarded as purely provisional.

Any further serious discussion of the design of de' Dondi's astrarium must await a detailed comparative study of the surviving manuscripts of his treatise, and of their provenance.[135] In particular, it would be of the utmost interest to identify or find the copy of de' Don-di's work which was in the ducal library at Milan, and which almost certainly accompanied the astrarium when it was placed in the library (see above, pp. 23, 27). It is, moreover, quite likely that further copies of this work will be found once scholars generally are alerted to its importance.[136] Not only should a study of the manu-scripts be made with the object of establishing a basic text and *variorum* edition, but a carefully annotated translation should be made which would include, as an indispensable auxiliary, reproductions of all the dia-grams, and such variations of these as occur in the several manuscripts. Nothing less is deserved by one of the earliest surviving works on mechanical horology.

Following is a descriptive list of the manuscripts now known. In compiling this list, it has been necessary to rely for the most part on such published descriptions as exist, since we have not been able to examine in detail all the manuscripts. For this reason we have omitted the incipits of the manuscripts and confined ourselves to such titles as we have been able to ascertain. Some of the information given will doubtless require revision.

Biblioteca Capitolare Vescovile, Padua

1. *Cod.D.39.* "Tractatus astrarii Johannis de Don-dis Paduani civis cui tres sunt partes." Probably writ-ten by an amanuensis under de' Dondi's guidance, and therefore before 1389 (the date of de' Dondi's death). Parchment, 34 folios and 2 larger guard sheets in 2 columns, size 36 × 25 cm. Drawings in color. Prob-ably a manuscript mentioned in the inventory of de' Dondi's personal property, filed by his widow (see above p. 14); it was given to the library in 1795 by

Monsignor Francesco Scipione, Marchese Dondi dall'-Orologio. Published in Giovanni Dondi dall'Orologio, *Tractatus astrarii. Biblioteca Capitolare di Padova, Cod. D.39, Introduzione, trascrizione e glossario a cura di Antonio Barzon, Enrico Morpurgo, Armand Petrucci. Giuseppe Francescato, con la riproduzione fotografica del codice* (Codices ex ecclesiasticis Italiae biblithecis selecti, phototypice expressi . . . vol. IX), Vatican City, 1960.

In referring to the manuscript copies of the *Tractatus astrarii* of Giovanni de' Dondi, his descendant, Mon-signor Francesco stated in his *Memoria* [89], p. 486, that he personally owned two different copies. He added that

. . . The first is a manuscript of considerable value [excel-lence] with sketches, and miniature paintings; it has the appearance of an archetype, but it does not seem to be in the customary handwriting of Giovanni Dondi. Its char-acter [lettering] is neither perfectly precise nor intelligible, as are almost all those of this century in the squaring of the words. The illustrations are not all equally well preserved, and while I was doubtful whether time alone was the cause of all their damage I had them all recopied with precision; some were commendably accomplished by Daniel Danieletti, a young man of great ability, and others by Ab. Cerati. The other copy is of the 1500's, incorrect and incomplete.

This source was quoted in 1824–1825 by Francesco Maria Colle. In his *Storia scientifico-letterari* [26], p. 187, he mentioned de' Dondi's treatise on the as-trarium

. . . of which manuscript copies exist in various libraries, in addition to that owned by Mons. Orologio, which could be the archetype, although it does not appear to be in the hand of [Giovanni de'] Dondi.

Biblioteca Nazionale Marciana, Venice

2. *Cod. 85, Cl. Lat. VIII, 17.* "Planetarium . . ." Fourteenth century; the date, M°CCC°LXXXXVIJ° (1397) appears across the cover sheet. Parchment, 43 folios in 2 columns, size 43 × 28 cm. Red initials. The folios are badly damaged in the upper corners. This manuscript lacks de' Dondi's introduction. A translation of this manuscript was used by Baillie [81], and Lloyd [43], pp. 9–24, & [97]. See Valentinelli [74], p. 262; Thorndike [117], *passim.*

Biblioteca Ambrosiana, Milan

3. *C. 221 inf.* "Opus planetarii Iohannis de Dondis fisici Paduani civis." Fourteenth century. Paper, bound in boards. 101 folios written in a Gothic hand in 2 columns, size 34 × 23 cm., with illustrations in color. On f. 101r – 101v is "Ordinatio quen feci berto de Padua pro communi horologo" (i.e., "Scheme for an ordinary clock which I, Berto of Padua, made"); *cf.* below, p. 42, no. 6. This manuscript, like the one fol-lowing. belonged to V. Pinelli (1535–1601). See Thorndike [117] *passim*, and Rivolta [62c], *passim.*

---

[135] For instance, it is known that at least two manuscripts of Giovanni de' Dondi's work remained in his family (see above, p. 14). The present location of only one of these is known (Biblioteca Capitolare Vescovile, Padua, MS D. 39).

[136] See the remarks on Henricus Ragnetensis' copy of de' Dondi's work, above p. 27 and below, p. 42, No. 8. There is also the possibility that there exists at least one other work describing in detail the astrarium, see above p. 28.

4. *C. 139 inf.* "Astronomia, Johannes de Dondis." Fifteenth century. Paper. 78 folios in single column, with illustrations usually unfinished and without captions. The de' Dondi section ends on f. 70ʳ, and is followed by figures, apparently unrelated to the astrarium, and by a German text and a Latin astronomical text, both in a later hand which, according to Thorndike, may be that of Pinelli (see above, no. 3) who owned this manuscript also. It is only in this later hand that de' Dondi is named as the author; the same hand has noted that this manuscript is neither so complete nor so fine as that described above as no. 3. Part of the binding consists of a draft in German of the will of a Thomas Benter, and the later additions to this manuscript might have some connection with him. See Thorndike [117], *passim,* Zinner [118], pp. 18–19, and Rivolta [62b], *passim.*

Biblioteca Civica, Padua

5. *CM 631. "Astrarium. . . ."* ("Esemplatum ab Antonio Lupatio A.D. 1466") stated elsewhere in the manuscript to have been completed by Lupatio on 7 November 1466. Vellum. See Zinner [118], pp. 18–20, and Parisi [102], pp. 17–18.

Bodleian Library, Oxford

6. *MS. Laud. Misc. 620.* "Opus planetarij Johanis de dondis fisici paduani ciuis." At the end, in the hand of the text of the manuscript is a statement, that it was written by John of Leyd ("Scriptum est per manum Johanis leyd") and, in another hand, that the drawings were completed, at the fourth hour on the 8 November 1461, by Jacomo Polito, stationer ("Desegnado per mi Jacome polito Cartolaro ad. 8. novembre. 1461, ore 4"). 101 folios in 2 columns size 34.5 × 23.5 cm. with illustrations in color. Preceding the colophon is a section headed "Ordinatio quam feci berto de Padua pro comuni horologij [sic]," which also occurs in the manuscript described above, p. 5, no. 3. This manuscript was given to the Bodleian Library in 1635, by Archbishop William Laud (1573–1645), as appears from a contemporary inscription on f. 1ʳ. Unfortunately, this is one of the Laudian manuscripts whose provenance is unknown (see Madan and Craster [46] pp. 15–18). See Coxe [27], pp. 446–447; parts of this manuscript have been reproduced by Baillie [81], Lloyd [43] and [97], and Morpurgo [101].

Wellcome Historical Medical Library, London

7. *MS 248* (Accession no. 41398). "Planetarium opus de fabrica horarii magestralis." Late fifteenth century. Paper. 83 folios in a single column, size 30 × 21 cm., with 59 drawings. This manuscript was acquired, for £8, at the public auction of the library of Sir Frank Crisp by Messrs. Sotheby, Wilkinson & Hodge, London (see [116]), where it was lot 94, sold on the first day of the sale, 17 November 1919. The sale catalogue described the manuscript as an "Astrological Manuscript of the early XVIth Century, written in Latin by an Italian scribe." Not surprisingly, it remained unknown to students of de' Dondi's work. It was accurately described, however, in the first volume of Moorat's *Catalogue* of western manuscripts in the Wellcome Library ([53], p. 152), published in 1962, and attention was first drawn to its existence in an article, published in 1963, by one of the present writers (Maddison [99], p. 33, n. 34).

Biblioteka Universytetu Jagiellońskiego, Cracow

8. *MS 577 (DD. III. 28).* Title added in different hand from text "Johannis de dondis fabrica horarij magistralis." At the end: "Τελως. Astrarij Johannis de dondis patauini Libellus explicit." Dated by Wisłocki as sixteenth century, but examination of a microfilm copy shows that the manuscript is clearly Italian, and most likely to have been written in the last quarter of the fifteenth century. Written in a fine humanist hand, the text and well-drawn illustrations carefully laid out on the pages; the text begins with a beautifully illuminated initial "C," the style of which suggests that the manuscript originated in Padua or the Veneto. 73 pages of paper with 22 blanks at end; 1 page of parchment. It would be of great interest to know how this manuscript came to Cracow (see above, p. 27). Perhaps served as original for no. 9 below. See Wisłocki, *Katalog* [78] p. 181, and Thorndike & Kibre [71], col. 1031.

9. *MS 589 (DD. IV. 4).* At end: "Τελως Astrarij Johannis de donis patauini Libellus explicit Scriptus per Henricum Ragnetensem [137] in studio vniuersali Cracoviensi. 1494." Paper. A few illustrations (some unfinished) at the beginning; blanks were left for the remainder, but these were never drawn in. This manuscript apparently forms part (pp. 445–496) of a volume which includes *i.a.,* Ptolemy's *Almagest,* and on p. 1 of which is the following inscription: Liber iste erat magrj Leonardj de Dopczicze [138] s. canonum baccalaurej, collegiatj maioris Collegij artistarum, datus pro libraria artistarum Studij Cracoviensis." The intimate connection of this manuscript with the University, and the circumstances in which it was copied there in 1494/5 are intriguing. The relative dates of the two manuscripts at Cracow, the identity of their explicits, and the unfinished state of the drawings in this manuscript, suggest that this manuscript may have been copied from the other (No. 8 above). See Wisłocki, *Katalog* [78], p. 187 and Thorndike & Kibre [71], col. 1031.

---

[137] It has not proved possible to identify Henricus Ragnetensis (i.e., of Ragnit, East Prussia).

[138] Leonardus a Dobczyce flourished *ca.* 1504. He is mentioned in an entry from the year 1504 in the records of the University; see Barycz, *Conclusiones Universitatis Cracoviensis* [8a], p. 99.

Eton College

10. *MS 172. Bi. a. 1. (Bo. 3.20).* "Iohannis de Dondis Physicus Paduanus de Conficiendis Horologiis omnium Planetarum." Fifteenth century. An Italian copy of a manuscript of 1397, probably Cod. 85 of the Biblioteca Nazionale Marciana in Venice (above, no. 2). Two volumes, as one. Part I consists of 58 folios of drawings; part II consists of 87 folios of text with additional drawings, and is entitled, "opus planetarii Joanis de Dondis fisici paduani ciuis." Very probably came to the College in 1639 from the Provost, Sir Henry Wotton (1568–1639). See James [39], p. 93.

Two other manuscripts of de' Dondi's work have been occasionally noted in various publications. One of these is said to have been owned by the library of Trinity College, Dublin. This statement is incorrect and, as noted by Parisi [102], p. 17, no such manuscript exists in that collection. The other manuscript was the following:

Biblioteca Nazionale, Turin

11. *MS. XLV.* "Traditur methodus construendi horologium a Johanne patavino inventum subjectis rotarum figuris." Fifteenth century. Vellum, 29 folios. This manuscript was lost in a fire which destroyed the library in January 1904. See Pasini [103], p. 475.

Finally, we must list a text in a manuscript, now in the library of the University of Salamanca, which though it makes no mention of de' Dondi, is clearly an incomplete version of his treatise:

Universidad de Salamanca

12. *MS. 2621,* item 12, ff. 25^r–76^v ["Opus manuale in quo motus omnes longitudinales planetarium oculo discernantur"]. Parts of the preface of this text, and the beginning of the first part (the whole text is not available to us) are almost identical with de' Dondi's treatise on his astrarium. The preface states that the work is divided into three parts, but the third part is missing, though ff. 77^r–80^r have been left blank for the insertion of this part. Manuscript Sal. 2621, "exceptionellement riche en textes inédits sur les instruments astronomiques du XVe siecle" (Beaujouan), contains thirty-one other items, including a *Compositio equatorii,* after Campanus' *Theorica planetarum* (see above, pp. 9–10, 16), a star-table of Jean Fusoris, a *Constructio et utilitates speculi planetarum* by Johannes Simonis de Zelandia, well-known works of Qustâ b. Lûqâ, Llobet of Barcelona, Prophatius Judaeus, *et al.,* and anonymous treatises on various instruments. The manuscript is of the fifteenth century, and Beaujouan has shown that it is of Dutch origin. It later belonged to Martin of Middleburg (Zeeland), and bears this inscription on f. 1^r: "Magister Martinus Medienburchensis, astrologus admirabilis, dono dedit, anno 1522, mense maio." It belonged to the Colegio de San Bartolomeo, Salamanca,

was subsequently at the Palacio Real, Madrid, and transferred to the University Library at Salamanca in 1954. The manuscript consists of 175 paper folios, and three of parchment, is written in two columns and is illustrated. For a full description see Beaujouan [12a] *Manuscrits scientifiques,* pp. 30, 60, and 163–173. Beaujouan, p. 167, quotes from the preface, the beginning of the first part, and the end of the second part of the *Opus manuale.* He remarks that "il s'agit d'un instrument imaginé en s'inspirant de l'équatoire attribúe à Campanus," and that "une comparaison attentive s'imposerait avec le 'Planetarium in tres partes distinctum' de Johannes de Dondis."

As mentioned above, p. 28, Magenta refers to an account of the de' Dondi *astrarium* written by Paolo Trizio. Magenta gives no source for this statement and we have been unable to trace this work or identify this author.

EIGHTEENTH–CENTURY DOCUMENTATION

The de' Dondi astrarium appears to have been ignored by scholars throughout the latter part of the seventeenth and early part of the eighteenth centuries. No related records or publications during this period have come to light. Fortunately, however, several illustrated manuscripts of de' Dondi's treatise on his astrarium have survived, and appeared sufficiently interesting for such bibliophiles as Archbishop Laud and Sir Henry Wotton to preserve in their libraries (see above, pp. 42, 43).

The astrarium was next described in a comprehensive study by C. Falconet which appeared in Paris in 1745.[139] Although this article incorporated numerous inaccuracies, it can be considered an important pioneer work. Falconet's article probably served as the source for the inaccurate entry for "Jacques de Dondis" in the Abbé de Fontenai's *Dictionnaire des Artistes* [34a] published in 1776 (I, p. 518). This dictionary records that Giovanni de' Dondi merely described a clock built by his father and accuses Regiomontanus of error concerning the true inventor of the astrarium.

Falconet's study was severely criticized in a work produced later in the century by a direct descendent of Giovanni de' Dondi, Monsignor Francesco Scipione, Marchese Dondi dall'Orologio, Bishop of Padua. The Monsignor's *Memoria* was first read before the members of the Accademia di Scienze, Lettere et Arti of Padua on 7 March 1782, and subsequently published in the *Atti* of the Accademia in 1789.[140] It was reprinted as a monograph in Padua in 1844 by Marchese Gaspare Dondi-Orologio, on the occasion of the wedding of his brother, Marchese Francesco Dondi-Orologio, to the Contessa Maria Nani of Venice.

---

[139] Falconet, "Dissertation" [92]. Read before the Académie des Belles-Lettres in June 1745.

[140] Mons. Francesco Scipione Marchese Dondi dall'Orologio, "Memoria" [89].

FIG. 39. Full front view showing horary dial of the model of the astrarium in the Museum of History and Technology, Smithsonian Institution, Washington. Photograph courtesy the Smithsonian Institution.

This *Memoria* is one of the most important and useful of the later sources on de' Dondi and the astrarium. It provides many valuable genealogical data from family archival records, as well as accurate renderings of earlier writings on the astrarium. It was the Monsignor's claim that the title of nobility "dall'Orologio" or "of the Clock" was awarded to Giovanni de' Dondi by Duke Gian Galeazzo Visconti II for his construction of the astrarium. Later writers have strongly contested this claim, however, asserting that the title was already being used by Jacopo de Dondi, the father, before his death and before the astrarium had been completed.

NINETEENTH-CENTURY REVIVAL OF INTEREST

Numerous works relating to the de' Dondi clocks, ranging from brief biographies to exhaustive studies of

contemporary documents, were produced in the nineteenth century.

The earliest of these, which includes a fairly comprehensive history of the family and summary of the works of Jacopo and Giovanni de' Dondi, was the *Storia scientifico-letterario dello Studio di Padova* by Francesco Maria Colle which was published in 1824.[141]

This was followed by a brief biographical sketch and bibliographical list in Giuseppe Vedova's *Biografia degli scrittori padovani* which appeared in 1832.[142]

The first book in English which considered the de' Dondi astrarium appears to have been the work of Johann Beckmann. His work appeared originally in German as *Beiträge zur Geschichte der Erfindungen* (1780–

---

[141] Colle, *Storia scientifico-letterario* [26] **3**: pp. 174–191.
[142] Vedova, *Biografia* [76], pp. 335–344.

FIG. 40. Full-length view showing details of perpetual calendar of the model of the astrarium in the Museum of History and Technology, Smithsonian Institution, Washington. Photograph courtesy the Smithsonian Institution.

FIG. 41. View of interior of movement of the model of the astrarium in the Museum of History and Technology, Smithsonian Institution. Photograph courtesy the Science Museum, London.

1805), and was subsequently published in English as *A History of Inventions, Discoveries, and Origins.*[143] Although the three sections on horology are relatively brief, Beckmann quoted some of the more important early sources on the astrarium, including Scardeonius and Savonarola.

Jacopo and Giovanni de' Dondi and the astrarium were again considered, although briefly, in *Biografia degli artisti padovani* by Napoleone Piertrucci which appeared in 1858.[144]

In 1860 an exhaustive study of the clock tower of St. Mark's in Venice was published by Niccolò Erizzo entitled *Relazione storico-critica della Torre dell'Orologio di S. Marco in Venezia.*[145] An interesting and useful account of the clocks of Jacopo and Giovanni de' Dondi was included in the first part of this work, as well as some consideration of the fate of the astrarium. Erizzo was among the first to assume that it was transported to Spain by the Emperor Charles V.

A study by L. T. Belgrano, "Degli antichi orologi pubblici d'Italia, Con aggiunta di notizie della posta in Genova," which was published in *Archivio storico italiano* in 1868[146] reviewed earlier sources and provided useful commentaries.

The biographies and works of Jacopo and Giovanni de' Dondi were comprehensively presented in several works by Andrea Gloria late in the nineteenth century. The first of these was *Monumenti della Università di Padova 1318–1405* which appeared in 1888.[147] This was followed by two long studies in the *Atti del Reale*

---

[143] Beckmann, *A History of Inventions* [13]. See n. 134.
[144] Piertrucci, *Biografia* [59], pp. 103–107.
[145] Erizzo, *Relazione* [34], pp. 57–62.
[146] Belgrano, "Degli antichi orologi" [83].
[147] Gloria, *Monumenti* [38] 1 : pp. 381–386.

FIG. 42. View showing details of the escapement of the model of the astrarium, in the Museum of History and Technology, Smithsonian Institution, Washington. Photograph courtesy the Smithsonian Institution.

Istituto Veneto di Scienze, Letteri ed Arti in 1896 and 1897.[148] These works related primarily to the clock of Jacopo de' Dondi, concerning which Gloria brought to light numerous contemporary documents, some of which also related to the de' Dondi family history.

Of equal importance is a volume by Vincenzo Bellemo entitled *Jacopo e Giovanni de Dondi dall'Orologio* which was published in Chioggia in 1894.[149] In this work Bellemo collected contemporary documentary records, many of which had not been previously considered, and presented critiques of other studies made in his own time, including those of Colle and Gloria.

TWENTIETH-CENTURY STUDIES

It was well past the turn of the century before the astrarium and the de' Dondi artisans came to scholarly notice again. The first work of the twentieth century on the subject was Vittorio Lazzarini's article on the inventory of the personal possessions of Giovanni de' Dondi,

---

[148] Gloria, "I due orologi" [94], pp. 675–736; and "L'Orologio" [95]. Reprinted separately in Venice, Tip. Ferrari, 1897.
[149] Bellemo. *Jacopo e Giovanni de Dondi* [14].

FIG. 43. The dial of the movable feasts, shown without its cover. From the model of the astrarium in the Museum of History and Technology, Smithsonian Institution, Washington. Photograph courtesy the Smithsonian Institution.

inherited by his heirs, entitled "I libri, gli argenti, le vesti di Giovanni Dondi dall'Orologio" published in 1925 in the *Bollettino del Museo Civico di Padova*.[150] It included interesting data about the manuscripts of Giovanni de' Dondi's own astrarium.

The first serious technological consideration of the astrarium was not made until 1934. This modern pioneer work was undertaken by the late Granville Hugh Baillie and presented in a lecture to the members of the British Horological Institute on 14 March 1934.[151] The lecture summarized his study of the subject which had been undertaken many years before. A brief account of the astrarium subsequently formed part of his valuable source book on *Clocks and Watches. An Historical Bibliography* which was published posthumously in 1951.[152]

The astrarium was also mentioned in Professor Lynn Thorndike's *History of Magic and Experimental Science* (Vols. III and IV, on the fourteenth and fifteenth centuries) which appeared in 1934.[153] In this

150 Lazzarini, "I libri" [96].
151 Baillie, "Giovanni de' Dondi" [81]. See n. 2.
152 Baillie, *Clocks and Watches* [5], pp. 1–2.
153 Thorndike, *A History of Magic* [70] 3, chap. XXIV; pp. 386–397, 740–741, 746.

work Professor Thorndike's consideration of de' Dondi was primarily in relation to his other fields of endeavor. Two years later, an important article by Professor Thorndike on "Milan Manuscripts of Giovanni de' Dondi's Astronomical Clock and of Jacopo de' Dondi's Discussion of Tides" in *Archeion. Archivio di Storia della Scienza* [117], undertook for the first time a comparison of some of the surviving manuscripts of the astrarium.

Within the past decade and a half, interest in the astrarium was awakened among the Italian horologists. In 1950 Enrico Morpurgo published his *Dizionario degli orologiai italiani*[154] which summarized most of the previously published references. This was followed by two articles in the Italian horological journal, *La Clessidra*. The first of these was by Bruno Parisi on the de' Dondi manuscripts describing the astrarium[155] and the second by Antonio Simoni on the astrarium itself.[156]

De' Dondi's masterpiece and some of the related manuscripts were described in Abbott Payson Usher's *A History of Mechanical Inventions* in 1929 and in the second edition which appeared in 1954, and in J. Drummond Robertson's important book, *The Evolution of Clockwork*, published in 1931.[157] This was followed by the first comprehensive study in the English lan-

154 Morpurgo, *Dizionario* [55]. See n. 128.
155 Parisi, "I Manoscritti" [102], pp. 17–18.
156 Simoni, "Giovanni de' Dondi" [114], *op. cit.* See n. 52.
157 Usher, *A History of Mechanical Inventions* [73], pp. 198–199. Robertson, *The Evolution of Clockwork* [62d], pp. 33–34.

FIG. 44. Detail of the horary dial of the astrarium, from the model in the Museum of History and Technology. Photograph courtesy the Smithsonian Institution.

FIG. 45. Dial of the sun, from the model of the astrarium in the Museum of History and Technology. Photograph courtesy the Smithsonian Institution.

guage, an article by H. Alan Lloyd on "Giovanni de Dondi's Horological Masterpiece 1364" which appeared in *La Suisse Horologère* in 1955.[158] This article, translated into Italian, was published in the horological journal, *La Clessidra,* in 1961. The same text was subsequently summarized in a chapter of Lloyd's later work entitled *Some Outstanding Clocks*

[158] Lloyd, "Horological Masterpiece" [97]. See n. 107. Idem., "Il Capolavoro d'orologeria" [98].

FIG. 46. Dial of Jupiter, from the model of the astrarium in the Museum of History and Technology. Photograph courtesy the Smithsonian Institution.

Over Seven Hundred Years 1250–1950 which appeared in 1958.[159] Lloyd has also included entries for Jacopo and Giovanni de' Dondi in his recent *Collector's Dictionary of Clocks.*

One of the most recent important contributions was the publication of a reproduction of the *Tractatus astrarii* (Cod. D. 39) owned by the Biblioteca Capitolare of Padua, which was undertaken by the Biblioteca Apostolica Vaticana in 1960.[160]

A recent addition to the literature of the astrarium is an article by Morpurgo entitled "Raffronto tra l'Astrario e il Planetario del Dondi" which appeared in *La Clessidra* in 1963.[161] The latest discussion of the astrarium in the context of early clocks is that in Ernest L. Edwards' *Weight-driven Chamber Clocks of the Middle Ages and Renaissance,* published in 1965.

FIG. 47. Dial of Mercury, from the model of the astrarium in the Museum of History and Technology. Photograph courtesy the Smithsonian Institution.

Although not specifically on the subject of the de' Dondi astrarium, several writings by Derek J. de Solla Price [162] which have been published in recent years need to be considered in connection with it. The first of these, entitled "Clockwork Before the Clock," appeared in the *Horological Journal* in December 1955 [106], and was followed by "The Prehistory of the Clock" in *Discovery* [108] in April 1956. A very important monograph "On the Origin of Clockwork, Perpetual Motion Devices and the Compass" was published in 1959 by the

[159] Lloyd, *Some Outstanding Clocks . . .* [42]. See n. 127. Lloyd, *Collector's Dictionary of Clocks* [43], pp. 70–72.
[160] Barzon *et al., Tractatus astrarii* [11]. See above, p. 41.
[161] Morpurgo, "Raffronto" [101]. Edwardes, *Weight-driven Chamber Clocks* [33a], *passim.*
[162] Price [106–108–109–110].

Smithsonian Institution [109]. An article entitled "Unworldly Mechanics" appeared in *Natural History Magazine* in 1962 [110]. A brief article on "Leonardo da Vinci and the Clock of Giovanni de' Dondi" was published in *Antiquarian Horology* in June 1958 [111] and has already been mentioned (see above, p. 31).

MODERN MODELS OF THE ASTRARIUM

It was inevitable that with the revival of interest in the mid-twentieth century, in the scientific and technological aspects of the astrarium, attempts would be made to reconstruct working models of the clockwork from the surviving manuscripts describing the astrarium.

The first working model was produced for the Smithsonian Institution in Washington and completed in 1960, a project initiated by H. Alan Lloyd.

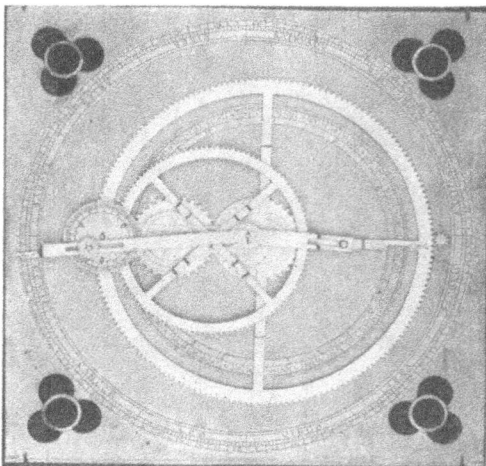

Fig. 48. Dial of Saturn, from the model of the astrarium in the Museum of History and Technology. Photograph courtesy the Smithsonian Institution.

A partial translation of Cod. 85 Cl. Lat. VIII, 17 (see above, No. 2) in the Biblioteca Nazionale Marciana, Venice had been made for the late G. H. Baillie, and subsequently acquired by C. B. Drover. This was used as a point of departure by H. Alan Lloyd for initiating his own studies which he published. Lloyd subsequently interested himself in having a reproduction produced of the de' Dondi astrarium. Toward this end, he attempted to interest English, Italian, and other museums in such a project, but without success. The only favorable reaction came from the Museum of History and Technology of the Smithsonian Institution in Washington, and the project was undertaken in 1958 through the support of Edwin A. Battison, Curator of Light Machinery and Horology. The reconstruction was

Fig. 49. Dial of the moon, from the model of the astrarium in the Museum of History and Technology. Photograph courtesy the Smithsonian Institution.

Fig. 50. Full-length view of the model of the astrarium constructed by Signor Luigi Pippa in Milan, now in the Museo Nazionale della Scienza e della Tecnica, Milan. Photograph courtesy Museo Nazionale della Scienza e della Tecnica.

FIG. 51. View of the astrarium by Signor Luigi Pippa showing details of the perpetual calendar, now in the Museo Nazionale della Scienza e della Tecnica, Milan. Photograph courtesy Museo Nazionale delle Scienza e della Tecnica.

built in England by Thwaites and Reed, Ltd., with the technical guidance of Lloyd. The actual construction was done by P. W. Haward, and the engraving by F. N. Fryer.[163] The foresight of the Smithsonian Institution has been amply rewarded by the considerable scholarly interest this first reconstruction of the astrarium has aroused throughout the scientific world.

More recently, in 1963, a second model of the astrarium was completed in Milan. This reconstruction was made by Luigi Pippa for Cav. Innocento Binda of the same city, who subsequently donated it to the Museo Nazionale della Scienza e della Tecnica, Milan, where it is presently on exhibit.[164]

Pippa's reconstruction was based for the most part on Cod. D. 39 of the Biblioteca Capitolare, Padua. (See above, no. 1.) Although Pippa departed from the text of the manuscript in some important details, the workmanship throughout, in the construction of all parts as well as in the engraving of the dials, is of the finest quality.

These two reconstructions of the astrarium may have initiated a museum trend, and may be followed by the production of other versions. This reawakening of interest in the astrarium, and consequently in the history of fourteenth-century technology, is most desirable and hopefully will lead to a fuller understanding and appreciation of the great tradition of mathematics and engineering which helped to produce the Renaissance.

---

[163] Anonymous, "The Dondi Clock" [80a].
[164] Curti, "Giovanni Dondi" [87], pp. 12–15.

APPENDIX

I. ORIGINAL TEXTS

Document I (de' Dondi, "Familia nostra . . ." [144])

Is Machinam perfectam suo ingenii acumine struxit, ubi multiplicata Orbium atque Planetarum congeries clare ac distincte moveri ordinate dignoschebatur, ut divinum potius quam humanum opus videretur.

Document II (Francesco Petrarca, "Testamento" [68])

Magistrum Johannem de Dundis, Physicum, Astronomorum facile Principem, dictum ab Horologio, propter illud admirandum Planetarii Opus ab eo confectum, quod Vulgus ignotum Horologium esse arbitratur. . . .

Document III (G. de' Dondi [32])

Quum Magnifici ac praepotentis domini domini Galeacij Vicecomitis Mediolani et comitis Virtutum filius praeclarae indolis in difficilem ac gravem morbum incidisset, integrum annum in Papiensi urbe ille assidere sum coactus.

Document IV (U. Decembrius [48])

Joannes de Horologio, Patavinus, astrologus suae aetatis acutissimus, cui Horologium illud, Princeps invictissime (Philippus Maria vicecomes), quod in tua celeberrima Bibliotheca situm est, cognomen imposuit.

Document V (P. de Mezieres [90])

Il est à savoir que, en Italie, y a aujourd'huy ung homme en philosophie, en médecine et en astronomie, en son degré singulier et solempnel, par commune renommée, excellent ès dessus trois sciences, de la cité de Pade. Son surnom est perdu : et est appelé maistre *Jehan des Horloges*, lequel demeure à présent avec le comte de Vertus, duquel pour science trebbe [triple] il a chacun an des gaiges et de bienfaits deux mille flourins ou environ. Cettuy maistre *Jehan des Horloges* a fait, de son temps, grandes œuvres ès trois sciences dessus touchiées, qui, par les clercs d'Italie, d''Allemagne et de Hongrie, sont authorisées et en grant réputation : entre lesquelles œuvres il a fait un instrument, par aucuns appelé Sphere, ou Orloge du mouvement du ciel : auquel instrument sont tous les mouvements des signes et des planettes avec leurs cercles et épicycles, et différences par multiplications, roes sans nombre, avec toutes leurs parties, et chacune planette en ladite sphère particulierement. Par telle nuit, on voit clairement en quel signe et degré les planettes sont et estoiles du ciel : et est faite si soubtilement cette sphère, que, nonobstant la multitude des roes, qui ne se pourroient nombrer bonnement sans défaire l'instrument, tout le mouvement d'icelle est gouverné par un tout seul contrepoids, qui est si grant merveille, que les solempnels astronomiens de lointaines regions viennent visiter en grant réverence ledit maistre Jehan et l'œuvre de ses mains ; et dient tous les grans clercs d'astronomie, de philosophie et de medecine, qu'il n'est memoire d'homme, per escrit ne autrement, que, en ce monde, ait fait si soubtil ne si solempnel instrument du mouvement du ciel, comme l'orloge dessusdite ; l'entendement soubtil dudit maistre Jehan, il, de ses propres mains, forgea ladite orloge, toute de laiton et cuivre, sans aide d'aucune autre personne, et ne fit autre chose en seize ans tout entiers, comme de a este informe l'escrivain de cettuy livre, qui a eu grand amitie audit maistre Jehan.

Document VI (Giovanni Manzini [41])

. . . Revidebam enim sphaerilogium tuis fabricatum manibus, deque tuae profundissimae mentis altitudine protractum in formam ; opus hercule ingentissimum, opusque divinae speculationis, opus denique hominum ingenio non dabile, neque operabile et nullis umquam diebus antelatis plasmatum : quamvis dicat Cicero, Possidonium sphaeram effecisse, cujus singulae conversiones efficiebant in sole et in luna, et quinque stellis errantibus, quod efficitur in coelo singulis diebus et noctibus. Non equidem reor tantam illi operi infuisse rationem artis, non id ingenii magisterium quod in isto. Nec per posterorum quemquam credo unquam factibile erit, vel saltem superabile ; cum non videamus in aetatum successione tam excelsas ingeniorum auctelas.— Sed quid fabor ? Decentius quippe foret hujusce rei silenium figere, quam exiguum loqui, ut de Carthagine Crispus inquit. In hoc enim opificio artificiorum plenissimo, tuis manibus elaborato, elimato, atque adeo mire sculpto, quod nullius artificis docta manus ipsum quiet agere, cernere datur certis discriminatos mentis septem stellarum, quae ab incessu dicuntur erraticae, circos sese orbiculata circinatione urgentes, quorum altior est is qui dicitur Saturni, quem ajunt in aethere minimum videri, sed ipsum maximo ambitu circumagi ; hicque tardivagus anno triceno ad metam a qua digressus fuerat regressum inchoat et orditur ; hic influentiam confert gelidam et rigentem. Post hunc Juppiter vertiginem suam rotat, volucriore tamen motu, annis scilicet duodecim complens cursum. Tertium Mars mox globum sortitur, bino peragens viam anno, cujus ardore protenso et superstantis pulpito primo senis rigore, qui falce recurva priscis temporibus pingebatur, contemperatum Jovem utrique interpositum salubrem influentiam dare constat. Sol autem lucis maximus lator solus omnium generabilium post Omnipotentem Deum productor et similiter propagator, medius existens perlucidissimis et amplissimis mentibus annuo per Zodiacum transitu complet orbem ; cujus ascensum pariter et descensum, temporum vices, mensium ordinem, signorum respectum, dierum augmenta, nec non decrementa, horarum denique spatia et momenta, et totum ejus officium subtilissimis figurationibus exemplorum per singulos gradus et puncta, perspicacissime designasti. Post hunc Veneris sidus, loca quod alterno perambulans itu ; nam Solis ortum praecurrens ut egregius praeco, primo Lucifer nuncupatur ; ab occidente vero lucem prorogans est nomen Vesperis consecutum. Hujus sideris influentia cuncta gignit in terris. Nam, ut inquit Plinius Veronensis, in alterutro exortu genitali rore conspergens, non modo terrae conceptus implet, verum animantium quoque omnium genitura stimulat. Signiferi autem ambitum peragit diebus illis totidem quibus Phaebus, ab ipso nunquam absistens partibus sex et quinquaginta longius, ut Timaeo libet. Penultimum situasti descensive Mercurium, illi proximum sed non ejus virium, non etiam quantitatis. Hic fertur novem diebus circulo ociore ; modo autem solis exortum, modo radians post occasum, ab eo nunquam XXII. partibus amotus, ut placuit Sofigeni. Phaebem, quam Lunam, Dianam, Dictynnam, Proserpinam et Cynthiam nostri prisci vates variis respectibus nominarunt. Planetarum ultimum, terrae magis proximum demum sistis ; quae nocturnarum tenebritudinum antidotum, variationis ambiguae, nunc exignam et inanem, nunc imbutam et bene pinguatam se offert ; nunc arcuatam cornibus, nunc circumflexam orbibus, nunc nitentem juventa, nunc languentem senecta, nunc rubore suffusam ventorum bajulam, nunc tinctam pallore nunciam pluviarum, nunc alba serenitate perlucidam tranquilli temporis quod a nobis mundo degentibus frequentius exulat hortatricem, quandoque latentem, quandoque serius, quandoque ocius apparentem; nonnunquam corolla praecinctam puri coloris aliquando, alias denique rutilantis: haec est aliquando sanguinea magnarum pertendictrix jacturarum, excelsa plagis in aetheris aliquando, contigua montibus aliquando; aliquo tempore vergitur erecta in boream, nonnunquam dejicitur versus austrum: ambitum quippe suum minimum obit; vicenio namque septenisque diebus, et tertia diei portiuncula tantundem aetheris emetitur, quantum gelidi senis sphaera trigesimo lustrat anno. Fratris ista fulgore suum radium occulit et assumit, amplificat ac extinguit; hujus ob interventum eumdem etiam Solem nobis occultari compertum est, ipsamque, Lunam terrae obice: radii enim Solis, Lunae interpositu opacati, splendorque Lunae terrae objectu, tenebras invehunt repentinas. Quae quidem omnia singulaque intuentibus perspiciaciter materiali exemplo vivo mirabiliter demonstrasti. Sed de his satis superque per me nescium siderum perstrictum sit; de quibus cum alio penitus siluissem, tecumque, ni fidens audacia de te sumpta me intimidum fecisses et promptum; cujus et non alterius in manum fore perventura, tam temeraiis ut sinerem non essem ausibus, quia possem facillime subsannari, ut in Oratore suo Cicero risum narrat ab Hannibale Phormionem, tanto tamque probato bellorum imperatori scientiae militaris monitus ingerentem. Porro circulus ille inferior in opere hoc tuo locatus in quo dies singula mensis cujusvis et totius anni computus et Kalendaria serietate noscuntur, et rimatae sese offerunt fenestrellae, festi solemnitatem, si quam ea luce celebrari contingat, ostendenti, pandentique cancellulo Sancti nomen designatum in circulo; non parum admirationis ignorantibus causam affert, intelligentibus vero delectationis ingentis. Atque ut paucis expediam, nullum magnis temporibus tam egregium, tam mirificum, tamque praestantis artificium ingenii non fuisse repertum.

Document VII (Registro delle bollette . . . Pavia [137])

Joannes de Clarii Magister ab orologiis deputatus as custodiendum orologium existens in castro magno Papiae pro ejus salario in mense L. 6.8.

Document VIII (P. C. Decembrio [31])

. . . ebbe nella sua Biblioteca in Pavia un insigne Orologio, sopra tutti quelli della nostra età memorabile, e quasi divino, fatto da Gio. da Padova, insigne Astronomo, in cui vedeansi i movimenti de' sette Pianeti.

Document IX (F. da Fabriano to Duke Francesco [119])

. . . Et oltra questo è bisogno acconciare li armarij che ruinano giù, acconciare certi banchi dove se sede chi vole studiare che sonno si stricti, male gli se pò intrare et questa è piccola cosa: ma sèra cosa bene facta bisogniaria conciare el coverchio dello Astrolabio che se tirasse su artifitiosamente con qualche girella, che stando come sta quando se leva, che non gli ha diligentia i più, el pone tra i libri et guasta el coverchio et i libri, et questo seria spesa de v libre imperiale. . . .

Document X (Missive Reg. 32 f. 17 [120])

Ser Fazino de Fabriano Cancellario.
Ne recordiamo già havere veduto uno libro in quella nostra libraria continente in che modo serà facto quello nostro horologio che è li et come si doveva fornire quello gli manca ad fornire qual libro già fu lassato in mane d'uno Magistro Benedicto nostro Medico, qual siamo certi lo remettesse poi al loco suo. Et perchè nostra intentione

è pur de fare fornire dicto horologio volimo che subito recevuta questa tu te retrovi col Bolognino et curi de trovare questo tale libro quale trovato volimo nel mandi qui subito. Mediolani die 29 aprilis 1456.

Marc. Jo.

Document XI (F. da Facino to Duke Francesco [121])

. . . Circha lastrolobio ovvero arrologio, credo la V. S. gli volglia procedere come bisognia: ma io non so si ha daltro luoco che de Milano maestro alchuno: e per che non havendo altri maestri che io sappia essere li in Milano et quando bene havesse de laltri ne uno qui per opera. Non dico per intendere i moti e i cursi de pianeti: ma per laltre cose gli sta lanimo fare come ad nisuno altra, et cosi credo fara come per experientia se potera vedere piacendo alla S. V. et credo che de opera non de fantasia melgliorera quello astrolobio assai: et la experientia sera quella el demostrarà che non creda la V. Ex. S. non gli habbia ancho facta su fantasia.

Appresso ad quisti di scripsi alla vostra Ill. S. alchune cose bisognavano in questa libraria, si per ornato si per necessità et alcuna per commodità, de che non ho havuto mai resposta: et non che scrivesse, ma le mandai in nota poi per d. Bartholomeo Trovamala che et come et la spesa seria come epso d. Bartholomeo de havere referito et da lui. Iterum se poteria intendere et questo et delli armarij che cascano, del coperchio dello Astrolabio che nel cavari et mettere et smettere guasta ogni cosa de i banchi da sedere. . . .

Document XII (M. Savonarola [66])

Neque in hoc Literarum genere parvum exstitit nostrae Urbis fidus Domus de Dondis, ab Horologio appellata, de qua post. Quae tot illustres Medicinae viros habuit, quorum doctrina & fama adhuc superexstat. Et in primis tertio loco Johannem ab Horologio aggredior, virum, ut intelliges, penitus divinum. Qui in Medicinae lectura omnipotenter valuit. Hic Orator magnus, his Medicus practicus, hic Mathematicus summus, hic manuum admirandus operator. Cujus sapientiam, doctrinam, ingenium quum Petrarcha Franciscus contemplaretur, & quadam sua in Epistola hujus admirandi viri dotes gloriosas discurreret, ait, nullum in Orbe doctiorem virum esse; & quod omnibus his detrahebat, servile & minime ejus liberale exercitum fuit. His Papiae Horologium suis manibus & ingenio fabricatum edidit: cujus admiranda est speciositas, in quo est firmamentum, & omnium Planetarum spherae, ut sic siderum omnium motus veluti in caelo comprehendantur. Festa edicta in dies monstrat, plurimaque alia oculis stupenda. Tantaque fuit ejus Horologii admiranda congeries, ut usque modo post ejus relictam lucem corrigere & pondera convenientia assignare sciverit Astrologus nemo. Verum de Francia nuper Astrologus & fabricator magnus fama Horologii tanti ductus Papiam venit, plurimisque diebus in rotas congregandas elaboravit. Tamdemque actum est, ut in unum eo, quo decebat, ordine composuerit, motumque, ut decet, dederit. Existimo quidem, mi Antoni, ipsum inter praeclaras Mundi res digne collocari, aggregarique debere. Res certe auditu stupenda, in seculo autem prius numquam audita, neque individuum visum tale. Omne enim individuum aliud par, aut quasi par reperiri contingat. Ab hoc quidem Horologio Domus gloriosa de Dondis cognomen accepit. Tanti autem viri ossa apud primam Cathedralis Ecclesiae portam arca quadam in autum elevata ornatu cum magno tenentur.

Document XIII (Missive Ducale [122])
Antonio de Tritio.
Perchè desideramo chel orologio in la libraria del nostro Castello de pavia se fornisca per le mane de maestro de

Guglielmo de parise, el quale e li in li serviti de la maesta del serenissimo Re, perche e le bono maestro et altra volta al tempo de la bona memoria del Ill. signore passato lo hebbe in le mane et comincio per fornirlo, e perche elle pur opus mirabile: pertanto vogli pregare la Maesta del Re, che la degni farvi comodita del dicto maestro saltem per quattro mesi, et potendolo havere, vogli mandarcelo qua da nuy piu presto che poray. Datum Mediolani die 12 maii 1456.

Cristof. C.

Document XIV (F. Malaguzzi Valeri [50])

Più tardi, in pieno secolo XV, il grande orologio non funzionava più e invano i duchi incaricaron meccanici e orologiai—Guglielmo Zelandino, maestro Claudio, un Zanino che già aveva costrutto uno strumento analogo per il castello di Milan (1478)—di accomodarlo. L'astrolabio fu tolto allora dalla biblioteca e mandato, nel 1494, in una sala del castello di Rosate.

Document XV (C. Simonetta to Bolognino de Attendolis [123])

Comiti Bolognino de Attendolis.

Essendoce pure continuamente all'animo quella nostra libreria, et deliberando nuj ponerli de l'altri libri, et augumentarli et mantenerli tucto quello gli havemo facto fare, havemo veduto et considerato bene, che non se attendendoli con diligentia, in pocho spatio de tempo se veneria ad nihilum de quello gli è facto, et intendando nuj ad qualche persona idonea et apta ad ciò, n'è parso quello magistro fece l'orologio in quello nostro castello assai apto et sufficiente ad attendere ad quello orologio et ad quella libraria, et provederli che 'l possa fare de conveniente salario, alla quale spesa ne pare per ogni modo gli debbiati contribuire vui per rata, secundo Fazino da Fabriano nostro cancellario, al quale havemo imposto de cio, ve dira et a luj crederite, in questa et dariteli fede pienamente come alla persona nostra. Mediolani, 28 maij 1459.

Cichus.

Document XVI (Registro delle Missive [124])

Scriptum fuit comiti Bolognino de Attendolis, quod ostendere deberet horologium, quod est in libraria, et omnes reliquies magistro Marco de Raynis de Mediolano. Datum Mediolani, die VI januarii 1460.

Aquilanus.

Document XVIa ([J. Regiomontanus] [62a])

. . . Quid multis moror? recentior occurrit Astronomus Ioanus de Dondis ciuis clarissimus, qui quantum & q; vere diuinum hoc studium coluerti atq; consecutus sit, reliquiae suae immortales docere coeperunt. Nunquid non memoratu dignum arbitraveris Astrarium eius, quod in arce Papiensi dux Mediolani depositum hodie tenet? cuius uidendi gratia Praelati ac Principes innumeri ad eum concessere locum, quasi miraculum quoddam spectari, & quidem non iniuria, tanta nempe & tam inusitata huiscemodi operis pulchritudo est atq; utilitas, ut nemo non possit admirari. ecce monumentum philosophi uestri nunq; periturum, ipsi iudices eritis, si studium Medicinae ciui uestro claritudinem peperit, an potius Astrorum peritia, utriq; enim tametsi plurimum operae impederit, illa quidam prosus tacetur, haec uero nomen suum immortale posteris effecti. Gaudere igitur o Patauini nobilissimi, quibus semper clarorum uirorum studia decori fuere. . . .

Document XVII (Missive N. 55 [125])

Abbati S. Faustini majoris brixie.

Havemo inteso quanto me ha referito petro ferrante nostro compare che la liberalitate vostra, quale ne vole usare in volerci compiacere del *libro Astrolabio* et il *quadrante* quale sapete ve habiamo facto richiedere, et perchè desideramo molto de havere dicte cose mandiamo li a voij a posta facta Bernardo de Corsicho nostro milanese presente exibitore, al quale piaciave consignare et dare dicte cose, le quale haveremo carissime et ne fareti cosa gratissima offerendone poij ad fare ad ogni vostro piacere maiore cosa. Mediolani die 26 Septembris 1463.

Cichus.

Document XVIII (Missive Registro N. 68, f. 31 [126])

. . . Et così sia vostra cura de rehaverli poi et de reponerli al loco suo, pretera ve remandiamo il libro delo horologio qual simelmente farete reponere al loco suo. Et de la repositione de questo et de la consignatione de li predicti ce darete avviso per vostre lettere. Mediolani 5 Novembris 1464.

Franciscus Sfortia Vicecomes manu propria subscripsit. Jo. Blan.

Document XIX (Registro Missive N. 98, f. 164 [127])

Domino Alexandro Sfortie.

Nuj havevamo deliberato de non lassare cavare fora dela libreria nostra de Pavia lo Virgilio, quale Vostra Signoria ne ha richesto in prestito per XX zorni. Ma havendo gustato el vino dulce, quale la ne ha mandato, havemo mutato el pensiero, et cosi dove non vi volevamo compiacere del dicto Virgilio, hora ultra questo anchora siamo contenti de compiacerve del Astrolabio, che è in la dicta libreria. Scrivimo aduncha al nostro castellano de Pavia per l'alligata, che daghi el Virgilio ad qualunche messo dela Signoria vostra. Viglevani, xiiii februarij 1471. Cichus.

Alexander.

Document XX (G. Trotto [145])

Finito ditti balli se fece restare li soni, che era circa hore XXIIIJ 1/2, et se de' principio a fare la representatione.

El Paradixo era facto a la similitudine di uno mezo ovo, el quale dal lato dentro era tutto messo a horo, con grandissimo numero de lume ricontro de stelle, con certi fessi dove steva tutti li sette pianiti, segondo el loro grado alti e bassi. A torno l'orlo de sopra del ditti mezo tondo era li XIJ signi, con certi lumi dentro dal vedro, che facevano un galante et bel vedere: nel quale Paradixo era molti canti et soni molto dolci et suavi.

Trete certi schioppi, et ad uno tratto cade zoso il panno de razo, che era dinanti al Paradixo, dinanti al quale remase uno sarzo fino a tanto che uno putino vestito a mo' de Angelo have annuntiato la ditta representatione. Livro de dire le parole, cade a terra ditto sarzo, et fu tanto si grande hornamento et splendore che parse vedere nel principio uno naturale paradixo, et così ne lo audito, per li suzvi soni et canti che v'erano dentro. Nel mezo del quale era Jove con li altri pianiti apreso, segondo el loro grado. Cantato et sonato che se have un pezo, se fece pore scilentio ad ogni cosa: et Jove con alchune acomodate et bone parole rengratio el summo Idio che li avesse conceduto de creare al mondo una così bella, legiadra, formosa et virtuosa donna come era la Ill.ma et ex.ma M.a duchesa Isabella. Una più bella et formosa creatura di lui; Giove li respose che non se ne doveva maravigliare perchè, quando lo creò lui, se reservo de potere creare una più bella et formosa creatura di lui, et che fin qui la haveva reservato per concederlo et donare a la Ex.ma M.a duchesa Isabella, et che voleva discendere in terra per exaltarla et gloriarla. Et così discese del Paradiso con tutti li altri pianiti, et andò in vetta de uno monte, et de grado ingrado ditti pianiti se li poseno a sedere apreso. Como furno tutti aseptati, mando per

Merchurio a noctificare a M.ª preditta, como era disceso in terra per honorarla et exaltarla et magnificarla et per donarli le tre gratie et acompagnarla da le sette vertù . . . et così Merchurio ando da sua ex., et con molte bone parole li noctificò la venuta de Giove in terra; et poi retornò a Giove la risposta. Audito questo li 6 pianiti, et inteso la raxone perche era venuto in terra, tutti a uno a uno rengratiorno Jove de la revelatione che li haveva factto de una tanto bella et virtuosa donna che aveva creato al mondo, confermandolo ne la sua volonta de doni li voleva fare, et zaschuno de loro, per hordine, li offerse la virtù. . . .

Document XXI (B. Bellincioni [15])

. . . che v'era fabricato, con il grand' ingegno et arte di maestro Leonardo Vinci fiorentino, il paradiso con tutti li sette pianeti che giravano, ed li pianeti erano reppresentati da homini, in forma et habito che si descriveno dalli poeti, li quali poeti tutti parlano in laude della prefata duchessa Isabella.

Document XXII (Leonardo da Vinci [142])

. . . Quando si apre il paradiso di Plutone alor sian diaboli che son in dodici olle, a uso di bocche infernali, quivi sia la morte, le furie, cenere, molti putti nudi che piangino, e vivi fochi fatti di vari colori. . . .

Document XXIII (Carteggio Generale 1494, 6 Nov. [128])

Gualterio de Basilicapetri.

Havendo inteso quello che ne hai facto significare per Johan Jacomo Gilino delo Astrolabio, quale è in la sala de Rosate, te dicemo lo lassi stare lì finchè saremo a Milano, perchè alhora ordinaremo quello vorremo se ne faccia.

Document XXIV (B. da Palude to Duke [134])

Ill.mo et Ex.mo unico Signore mio.—La Ex.ª V.ª di nuovo intendrà come passano li laurerii del castello. La camera nova che fa depinzere Bramante che è apreso de la strada è finita de refare de zeso [gesso]. La camera che è apreso de la capella se depinzerò. Quella del cello tondo non è metuto anchora ordine lavorare. Bramante è andato a Pavia per tore alcune cosse dal fiolo de maistro Ambrosio da Rosa per potere fare lavorare a dicta camera. La gronda de legname de dicte camere è fornita et coverta. La camera et la guarda camera de la Ill.ma Ducissa questa septimana serano fornite . . . La salla de sotto de dicta loza è finita de intonigare et è principata a fare il solo de madoni. La camera de dicta salla è intonigata et facto il solo . . . A V.ª Ill.ma S.ia de continuo me racomando. Viglevano ex acre 4 Marcii 1495.

Servitor Fidelissimus Blanchinus de Palude.

Document XXV (Registro Ducale N. 121, C. 98 [131])

Ill. et Ex. signor mio. È stato qua Bramanti ingeniero de v. Ex. quali dice havere comissione di quella de cavare alcuni desegni ne lo orologio che è in questa libreria de certe pianeti per ornare uno certo celi de una camera ad Vigiveni, et io per non havere altra commissione dela Ex. V. non lassarò exportare fora dela dicta libraria designo alcuno sino che non habia speciale licentia de quella. Sichè gli piaccia darne adviso de quanto haverò ad fare per littere signate de sua propria mane. A la prelibata v. Ex. de continuo me ricomando. Ex Arce Papie, die 5 Martij 1495.
E. Ex. V.

Fidelissimus servitor

Jacobus De Pusterla ibidem Casteltanus etc. [On reverse:] Ill. et Ex. Principi domino domino suo singularis-

simo Domino Ludovico Mariae Sfortiae Vicecomiti duci Mediolani etc.—Cito, Cito. [Obliquely:] Tristano qui faciat literas quas requirit.

Document XXVI (Registro Ducale, Libro 193, f. 211 [132])

Castellano Papie.

Siamo contenti lassiati pigliare a Bramante nostro ingegnero quelli designi del Horologio di quella liberia che a lui parira, et cossi ve ne facemo, libera licentia con questa. Mediolani, 6 martii 1495.

Document XXVII (B. Scardeone [67])

Praecipuum quoque patriae decus, & aevi sui fuit Ionnes Dondus illustris philosophus, & medicus, & mathematicus praestantissimus: genere quidem & origine Pat. ut pater, casu tamen Clodie natus quod tunc forte ibi Iacobus pater in ea urbe medicum ageret. Extat huius praeclarissimum opus, tribus voluminibus distinctum, lineisque & figuris mirabili arte signatum, ad conficiendum planetarium: ubi cursus caelestium signorum, libratis ponderibus facile metiantur, sicuti horarum: ut iam antea pater eius fecerat in Horologio suo. Vidi primam partem apud Galeatium Horologium, civem praestiss. Huius operis titulus est, PLANETARIVM IOANNIS DE DONDIS CIVIS PADVANI. Huius operis tres sunt tomi, in quorum primo est doctrina componendi planetarium, in quo capita numerantur XXV. Reliquos duos extare audio apud alios eiusdem familae alumnos. Iste autem Ioannes, Francisci Petrarche amicissimus fuit. . . .

Document XXVIII (H. Cardanus [22])

Ianellus Turrianus Cremonensis, cuius etiã supra meminimus, vir acris ingenij multa talia aut excogitauit, aut ab aliis excogitata in melius traduxit. Quemadmodum machinam illam mundi vniuersalem olim a Gulielmo Zelandino fabricatum atque dissolutam, in tenebrisque per incuriam marcescentem, cum ego quodam bono fato ad instaurandum bonas artes etiam obiter non minus quam ex industria natus, in luce reuocassem, Ianellus in integrũ eam restituit. Cuius exemplo alia Carolo Quinto Caesari ita construxit, vt in ea & temporum momenta, & partes signorũ singulas videas, & octaui orbis tardissimum motum intuearis. Diuisiones quoq; orbis signorum varias, quas domos vocant, horasq; aequales, & inaequales, & quod maius est, vniuersi orbis partibus inseruientes, vt haec vere machina orbem vniuersum referat, in ea inspicere licet. Omitto progessus regressusque singulorum errantium siderum, latitudines altitudinesq; aliaque innumera, vt prorsus res non minus fama quam fide maior sit.

Document XXIX (B. Saccus [64])

. . . Dominante deinde in Transpadanis Ioanne Galeacio Vicecomite, fabricatum fertur eiusmodi horologium, non solum horarum, sed etiam syderum expressis notis, atq; temporibus: ac Solis, Lunaeq; meatibus: cuius operis autor ignoratur: collocatumq; illud horologium in arce, vel castello Papiae fuit, vbi defuncto principe tam mirabile opus despectum iacuit, circulis etiam a suo loco sublatis. Exacto postea saeculo anni millesimi & quingentesimi, circa annum vigesimum nonum, quo CAROLVS QVINTVS Bononiae Imperialem coronam suscepit, allatum eidem Imperatori illud horologium incompositum (vt erat) situ, ac rubigine foedatum, fuit: quo conspecto, machinam admiratus, curari tanti operis instaurationem fabris vndiq; euocatis iussit. Quibus circa opificii restitutione frustra laborantibus, unus accessit Ioannes Cremonensis, cognometo Ianellus, aspectu informis, sed ingenio clarus: qui tantum opus speculatus, refici posse machinam dixit: sed nequiquam profuturam, ferris rubigine atritis, exesisq; nisi nouum instrumentum ad illius vetisti similitudinem, ac symmetriam componatur. Aggressusq; opus, siue priorem artificem immitando, atq; aemulando, siue exaequando diuturno labore opificium absoluit: quod deferri in Hispania Imperator voluit, magristro Ianello simul deducto. . . .

Document XXX (S. Breventano [19])

Haveva questo palagio quattro torrioni, ma hora non cene sono se non duoi nella facciata verso la Città, che quelli duoi che rimiravano verso il Parco furono (come habbiamo detto) gittati a terra da Lotreco Guascone cō l'artigliaria, sopra quello che nel entrar in detto Castello resta alla man destra era a giorni miei vn horologio di maravigliosa fattura, gia fatto fare dal Duca Giovanni Galeazzo Visconte, il quale nō solamente col segno e col suono della cāpana dimostraua l'hore, ma etian dio tutti i corsi & il girare de pianeti & segni celesti Questo per le mutationi dello stato nō essendone hauuto cura, corroso dalla rugine & leuate le ruote da i luoghi loro, ando tutto in ruina, e raccolte poi da vn Maestro Gianello Cremonese huomo di acutissimo ingegno in cotal arte ad instanza di Carol Quinto Imperadore a quella somigliāza ne fabrico vn' altro.

Document XXXI (G. B. Pietragrassa [60])

Resto spianata (dall'artiglieria di Lautrec) gran parte della muraglia, et le due torri parimenti del castello che risguardavano nel parco, in una delle quali si conservava la libreria. . . .

In un'altra delle torri suddette fu posto un orologio, qual era per meraviglioso artificio et per inestimabile fattura graziosissimo, i cui fragmenti per l'eta lunga dispersi, et per la ruggine consumati, Tonello, cremonese, architetto di Carlo V Cesare et geometra molto reputato, raccolse et imito cotanto bene, aggiungendovi ancora di piu che non aveva. In arte ed in industria supero poi quello con li suoi che fece; fra quelli uno quasi microcosmo fu riputato cosa sopra umana. ed in Toledo da Filippo, re cattalico di Spagna, secondo figlio del medesimo Cesare, fu conservato, ed in gran stima tenuto, in cui i moti di cieli cosi agevolmente si scoprivano, che avea del sovrumano piu che altrimenti.

II. BIOGRAPHICAL INDEX

All proper names which have appeared in the main text and in the notes have been assembled here in alphabetical order for the reader's convenience in making identification quickly. Where it has been possible to do so, additional names in the notes have also been added.

Adorno, Antonio, Doge of Genoa and personal friend of Giovanni de' Dondi, who visited him in June, 1389 just prior to his final illness.

Ambrogio of Rosate. See *Varese, Ambrogio.*

Attendolis, Conte Bolognino de, "comiti sancti Angeli et castelano castri papie," was the custodian of the castle of Pavia for Duke Francesco Sforza in the mid-fifteenth century.

Azzone (Visconti), son of Duke Galeazzo Visconti, whom de' Dondi attended in 1378–1379 during his illness.

Bellincioni, Bernardo, one of the minor fifteenth-century poets. A native of Florence, he was called to Milan to become court poet of Lodovico il Moro. Although he was Lodovico's court favorite, he was greatly disliked by Isabella of Aragon and by Leonardo da Vinci.

Benedict, Master, personal physician of Duke Francesco Sforza, noted in latter's letter of 29 April, 1456.

Benter, Thomas, German, fifteenth century. His name occurs as the maker of a draft of a will in German which forms part of the binding of MS. C. 139 inf. of the Biblioteca Ambrosiana. An appendix in German which forms part of the same manuscript might perhaps also be by Benter. Has not been otherwise identified.

Berto, of Padua, noted in MS C. 221 inf. of the Biblioteca Ambrosiana and in MS. Laud. Misc. 620 of the Bodleian Library, as the maker of "a scheme for an ordinary clock." Has not been otherwise identified.

Bescapé, Gualtiero da, attaché of the ducal chancery of Lodovico il Moro.

Bramante, (Donato d'Agnolo) (1444–1514), Italian Renaissance architect who developed a particular style of architecture called "Bramantesque." He worked at Milan between 1472 and 1499 for the Sforza dukes and then in Rome for Popes Alexander VI and Julius II until his death. He was assigned by Lodovico il Moro in 1492 to remodel the castle and square at Vigevano.

Breventano, Stefano, sixteenth-century Italian writer and author of *Istoria della antichità, nobilità, et delle cose notabili della città di Pavia* . . . (Pavia, 1570).

Bylica, Marcin (1433–1493), native of Olkusz, was a noted Polish astronomer. He was a pupil of Andreas Grzymala of Poznań, and studied at the Jagellonian University at Cracow. He first met Regiomontanus in Rome in 1464. In 1466 he accompanied him to Pressburg in Hungary (now Bratislava, Czechoslovakia), where they both remained to teach at the University. After Regiomontanus returned to Nuremberg, Bylica remained in Hungary and was appointed court astrologer to King Matthias Corvinus at Buda. He accompanied the king on all his campaigns. His collections of astronomical instruments were willed to the University of Cracow and deposited there in 1494.

Cajmi, Franchino, tutor of the children of Duke Francesco Sforza in 1464.

Calze, Giovanna di Reprandino dalle, first wife of Giovanni de' Dondi, married September, 1354. Mother of a son, Jacopo, and four daughters.

Campanus, Johannes, thirteenth-century mathematician and astronomer, born at Novara. Author of a *Computus* and a *Theorica planetarum.* The latter work was used by Giovanni de' Dondi as a basis for the design of the astrarium (see p. 16).

Cardano, Girolamo (Lat., Hieronymus Cardanus) (1501–1576), was a noted Italian mathematician, physician, and astrologer who was born in Padua. After graduating in medicine from the University of Padua he became a public lecturer in geometry at Milan in 1534, and was appointed professor of medicine at the University of Pavia in 1559 and at Bologna in 1562. He was arrested for heresy or debt in 1570. Upon his release he moved to Rome, where he was subsequently pensioned by Pope Gregory XIII. Cardano was the author of *Ars Magna* (1545) and *De Subtiltate* which was first published in Nuremberg in 1551, followed by *De Rerum Varietate* (1557). He mentioned the astrarium in *De Subtiltate* and claimed that he had in fact brought it to light again (see p. 37).

Charles V (1500–1558), King of Spain as Charles I (1516–1556) and Holy Roman Emperor (1519–1556). Son of Philip of Burgundy. Defeated and imprisoned the Pope in 1527, and was crowned King of Lombardy in 1530. In 1557 he retired to the monastery of San Yuste in Estremadura, following his abdication.

Clarii, Joannes de, custodian of the tower clock of the castle of Pavia and of the astrarium in 1399.

Claudio, a master clockmaker who worked in Milan in the mid-fifteenth century, and was banished from the Dukedom of Milan in 1475. May have been one of the clockmakers who attempted the repair of the astrarium.

Decembrio, Pier Candido (1399–1477), born at Pavia, became secretary to the Duke of Milan in 1447. He was appointed apostolic secretary to Rome in 1453, and later lived for some time in Naples before his death in Milan. Stated to have been the author of more than 127 treatises. Son of Uberto Decembrio. Wrote a biography of Philippo Maria Visconti, which mentioned the astrarium.

Decembrio, Uberto, author of *De Republica,* (a manuscript in the Biblioteca Ambrosiana) and father of Pier Candido Decembrio.

Dorn, Hans (*ca.* 1430–1509) of Vienna, famous maker of astronomical instruments who constructed the globe and other instruments of Marcin Bylica. He studied astronomy in Vienna under Georg Peuerbach, and went to Buda in 1476 to continue constructing instruments for the observatory, a work which had been interrupted by the departure of Regiomontanus for Nuremberg five years earlier. Dorn must have met Bylica at Buda and received the order for the construction of the globe. In 1478 King Matthias sent Dorn to Nuremberg to purchase books and instruments from the estate of Regiomontanus. All of the belongings of his former colleague, however, had been purchased by Bernhard Walther, and after a sojourn of six months in Nuremberg, Dorn returned to Buda apparently without any of Regiomontanus' instruments or books. After the death of King Matthias, Dorn returned to Vienna, where he worked in a Dominican monastery until his death in 1509.

Fabriano, Fazino (Facino) da, chancellor of Duke Francesco Sforza, who may have also served in the capacity of his librarian at Pavia in 1459.

Fondulo, Giorgio, of Cremona, perhaps served as tutor of Gianello Torriano *ca.* 1525–1535. (See below, s.v. *Gianello Torriano*).

Gaspare d'Allemagna, a man whose name appears on the rolls of the persons in the service of Duke Galeazzo Maria Sforza in 1467, as "custos relogii" or "custodian of the clocks" of the castle at Pavia. In 1470 he was no longer court clockmaker. No further identification of this German has been possible.

Gilino, Giovanni Jacomo, notified Lodovico il Moro on behalf of Gualtiero de Bescapé that the astrarium was at the castle of Rosate in November, 1494.

Gilliszoon, Willem, called also of Carpentras, Willelmus Aegidii de Wissekerke and Guillelmus Zelandinus. First records of his work are noted in 1476, when he furnished King René I of Sicily with a number of gnomonic and astronomical instruments. He was the author of a treatise on an equatorium in 1494, and may have been the clockmaker at the French court summoned to Pavia to repair the astrarium before 1440 and again in 1456.

Inverardi, Colonello Ludovico, military engineer who supervised the reconstruction of the castello Visconteo-Sforzesca at Vigevano between 1854 and 1857.

Isabella of Aragon (1470–1524), daughter of Alfonso II, King of Naples, who married Duke Gian Galeazzo Sforza on 13 January, 1390.

John of Leyd was the scribe who produced the fifteenth-century copy of the *Opus planetarium* of de' Dondi (Laud Misc. 620) in the Bodleian Library, Oxford.

Lebeuf, Abbé, eighteenth-century writer who compiled the *Histoire de l'Academie royale des Inscriptions, et Belles-Lettres, tirés des registres de cette Académie depuis l'année MDCCXLI jusques et compris l'année MDCC-XLIII.* Under the title "Notice des Ouvrages de Philippe de Maisieres, Conseiller du Roi Charles V, et Chancelier du royaume de Chypre" in Tome XVI, pp.

227–228 (Paris, 1751), he included the "Songe du vieil pèlerin" which described the astrarium.

Leland (or Leyland), **John** (1506?–1552), noted English antiquary, educated at St. Paul's School, London and at Cambridge University. After further studies in Paris, he took holy orders and became librarian to Henry VII prior to 1530, and was appointed king's antiquary in 1533. He toured England 1534–1543 for the purpose of collecting data for his proposed history of antiquities of England which was never published. His researches were described in his *A New Year's Gift* published in 1545. Leland became insane in 1550, and died two years later. His *Leland's Itinerary* was published first in 1710 at Oxford and his *Collectanea* in 1715.

Leonardus a Dobczyce (fl. *ca.* 1504), owner of a manuscript copy by Henricus Ragnetensis of de' Dondi's treatise on the astrarium, which may in fact have been made by Henricus for him.

Lupatio, Antonio (fl. 1466), noted in MS. CM 631 of the Biblioteca Civica at Padua as having completed copying that manuscript on de' Dondi's manuscript on the astrarium on 7 November 1466.

Maisieres, Philippe de, 14th-century writer and author of "Songe du vieil pèlerin" which was written between 1383 and 1388 and which described the astrarium. He served as adviser to King Charles V of France and as chancellor of King Peter of Cyprus. In 1365 he obtained Venetian citizenship, and it was probably in this year that he traveled to Padua where he saw the astrarium. He was a personal friend of Francesco Petrarca and of Giovanni de' Dondi.

Manzini, Giovanni, born in Villa Motta near Fivizzano, studied in his youth first at Sarzana, then at Parma. He later spent seven years at the school of jurisprudence at Bologna University. He went to Pavia late in 1387 or early in 1388 and served with the army of Gian Galeazzo Visconti against Antonio Scaliger, Signore of Verona. He took up his studies again upon his return to Pavia, and tutored the children of Pasquino Capello. He became a legal counsellor of note, interested in belles-lettres with a substantial library of his own. His friends included numerous important figures such as Giovanni de Traversi, Spinetta Malaspina, Andriolo Occhi of Brescia, Francesco Casini, physician to Pope Urban VI, Jacopo dal Verme, and Giovanni de' Dondi. In 1405 he was appointed mayor and captain of the Republic of Pisa, in which capacities he served with great honor. He died before 1422. Thirteen of Giovanni Manzini's letters in Latin were published by the Abbate Lazzari in his work entitled *Miscellaneorum ex manuscriptis libris Bibliothecae Collegi Romani societatis Jesu* (Rome, 1854).

Müller, Johann. (See *Regiomontanus*).

Palude, Bianchino da, architect of Lodovico il Moro for the castle at Vigevano.

Petrarch, Francesco (1304–1374), a noted Italian poet and scholar, was born at Arezzo and educated at Avignon, Montpelier, and Bologna. He met Laura, the inspiration of his *Rime* while he was living in Avignon *ca.* 1327, and thereafter devoted himself to the study of the classics. Was crowned poet laureate in Rome in 1341 and settled in Milan in 1353, where he served on diplomatic missions for the Visconti dukes, of whom he became a protégé. Among his friends were Giovanni Boccaccio and Giovanni de' Dondi. In his last will and testament, he left a sum of money to de' Dondi for the purpose of purchasing a gold ring with which to remember their friendship.

Pietragrassa, Giovanni Battista, historian who in 1636 wrote a manuscript history of Pavia from its early origins. This manuscript, in the library of the University of Pavia, was translated from Latin into Italian and published in 1760 by Leopoldo Arena.

Pusterla, Giacomo da, custodian of the castle at Pavia for Lodovico il Moro, as Duke of Milan, in 1495.

Pusterla, Paola da, brother of Giacomo da Pusterla and also employed at the castle at Pavia in 1495.

Ragnetensis, Henricus, In 1494 he copied in Cracow a manuscript of de' Dondi's astrarium (MS 589 (DD.IV.4) now in the Library of the Jagellonian University in Cracow). The copy may have been made to the order of Leonardus a Dobczyce, its subsequent owner.

Regiomontanus (1436–1476), was born in Königsberg. This German mathematician and astronomer established an astronomical observatory, printing shop, and workshop for making scientific instruments in Nuremberg with the patronage of Bernhard Walther, a wealthy patrician. In 1473 Regiomontanus published *Ephemerides ab Anno 1475–1506,* which were widely used. He substantially advanced mathematical and astronomical studies in Europe, and was called to Rome by Pope Sixtus IV to consult in the reformation of the calendar. He personally inspected the de' Dondi astrarium when he visited Pavia in 1463 and mentioned it in a lecture at the University of Padua. He subsequently devised an eight-sided astronomical clock based on de' Dondi's masterpiece.

René I (1409–1480), Duke of Anjou, Lorraine, and Bar; Count of Provence and Piedmont; and King of Naples, Sicily, and Jerusalem. René was well known as an amateur of the arts, particularly of poetry and painting. Willem Gilliszoon furnished him with scientific instruments in 1476. In September, 1453, René I visited Pavia, where he was received with great honors. He toured the city and visited the Castello Visconteo including the ducal library, where he presumably saw the astrarium. On the instructions of Duke Francesco Sforza, René I was presented with the keys of the castello, which pleased him greatly.

Richard of Wallingford (1292?–1336), Abbot of St. Albans. His father was a blacksmith of Wallingford, who died when Richard was ten. The boy was adopted by the Prior of Wallingford, who had him enrolled at Oxford. He was admitted to the abbey of St. Albans six years later. After training for three years, he returned to Oxford. In 1326 he was elected abbot of St. Albans. He succeeded in reducing the debt of the almost bankrupt abbey, and rebuilt its cloisters. The two portraits of Richard which have survived exhibit the spots on his face from the ravages of leprosy, which he had contracted at Avignon. His manuscript works include instructions for the construction of two instruments of his invention, the rectangulus and another which he called the "Albion" and of his astronomical clock, installed in the abbey of St. Albans, *ca.* 1327–1330.

Rosate. See *Varese, Ambrogio.*

Sacco, Bernardo (1498–1579), was born in Pavia, the son of Giacomo Filippo Sacco, who had been nominated by Francesco Sforza II as president of the Senate. Bernardo studied law but did not complete his doctorate. During the occupation of Pavia he stayed at Mirandola, and reentered the city after it had been sacked in 1527. Sacco devoted himself to reassembling and preserving the scattered records of the archespiscopal archives. In 1534 Francesco II sent Sacco to France on a mission to King Francis I. After his return to his city in 1535, he went to Rome on behalf of the Marchesa Virginia Gambara. There he was patronized and praised by Popes Paul III and Julius III. After living for nine years in Rome he returned once more to Pavia, where he died in 1579. He was the author of *De Italicarum rerum*

varietate et elegantia which was published in Pavia in 1565, in which he mentioned de' Dondi's astrarium in Lib. VII, Cap. XVII. Sacco's life is described by Pietro Terenzio in "Notizie della vita e delle opere di Bernardo Sacco pavese" in the *Manuale della Provincia di Pavia* (Pavia, 1857).

Scardeone, Bernardino (Bernardinus Scardeonius) (1478–1574), Italian writer and author of *De antiquitate urbis Patavii & claris civibus Patavinis* . . . which was published in Basle in 1560, and in which he includes an account of de' Dondi's scholarly work.

Sforza, Duke Alessandro, uncle of Duke Galeazzo Maria Sforza. He accumulated an important library which included Greek works translated into Latin, as well as a fine collection of works on astrology, medicine, cosmography, history, and poetry, which he collected from all parts of Italy. The collection was greatly increased by his son, Costanzo. His biography is in the works of Vespasiano da Bisticci. He died on 3 April, 1475.

Sforza, Francesco (1401–1466), Condottiere of Milan and son-in-law and successor of the last of the Visconti dukes, Filippo Maria. Considered to have epitomized the triumph of genius and individual power among fifteenth-century rulers.

Simonetta, Cicco (*ca.* 1400–1480), adviser and chancellor of Duke Francesco Sforza and of his widow, Duchess Bona of Savoy. He was also the father-in-law of Lodovico il Moro. On October 10, 1480, Simonetta, then of advanced age, was smuggled out of Milan by Lodovico in a cask and beheaded at dawn three days later in a meadow in Pavia.

Suso (or **Seuse**), **Heinrich** (?1300–1366), mystic, author of the *Horloge de Sapience,* was born at Überlingen on Lake Constance; his father's name was von Berg. Educated at Constance, Cologne, and Ulm; influenced by the work of Meister Eckhart. Suso's *Das Büchlein der ewigen Wahrheit,* written in Constance, discussed practical aspects of mysticism. What is now known to be a separate work of his, (and not as previously thought, the *Büchlein*) was translated from Latin into French in 1389 as the *Horloge de Sapience* by an anonymous French *maître de théologie,* a native of Lorraine and a member of the convent of the Observance at Neufchâtel. The original Latin text of the *Horloge* must have been written between 1333 and 1341 as, in the prologue, the author offers his book to the French *Maître général* of the *Frères Prêcheurs,* Hue de Vaucemain, whose term of office extended over that period. The *Horloge* is an account of Suso's mystical experiences and is primarily a dialogue between *Sapientia* (Wisdom) and her Disciple (the author); it also contains criticisms of disorder and deterioration within the Church and in the study of theology in the universities. The number of German, Latin and French manuscripts reflects the popularity of the work in the Dominican Reform Movement. The "clock" of the title and the prologue is an emblem of the soul and body of man which require supervision and regulation to function properly. It also reflects a fourteeth-century concern with the nature of time. Certain illustrations in manuscripts of the *Horloge* are of great interest to historians of horology as they depict various contemporary clocks, often with good detail. Much of the information in this note derives from Spencer, "L'Horloge de Sapience" [116*a*], which should be consulted for further information and useful references; see also Michel, "L'Horloge de Sapience" [100].

Torriano, Gianello (*ca.* 1500/15–1583/85), born at Cremona. Master clockmaker commissioned by Emperor Charles V to restore the astrarium in 1530. He subsequently entered the service of the Emperor and retired with him to the monastery of San Yuste in Estremadura after his abdication. It is believed that he died at Toledo.

Gianello Torriano remains a figure of mystery in the annals of horology and of technology of the sixteenth century. Even his name is not known with certainty. His family name is not a matter of record. It is believed that his cognomen was borrowed from the Torrazzo which symbolized his native city of Cremona. The date of his birth is given variously as the first decade of the sixteenth century, and as 1515. His parents were probably extremely poor, and incapable of providing for young Gianello's education. He was traditionally stated to have had a natural affinity for mathematics and the exact sciences, and to have studied for a time with Giorgio Fondulo of Cremona.

Torriano apparently first came to public notice, when the Emperor Charles V was in Italy for his coronation at Bologna in 1530. The Emperor was seeking a skilled clockmaker to repair the de' Dondi astrarium. Some sources state that Torriano was presented to the Emperor by Ferdinando Gonzaga, governor of Milan. Still another source relates that Torriano was brought to the Emperor's attention by Don Alonso Dovales, Marqués del Vasto.

It has been said that the Emperor advertised publicly, by means of a printed edict distributed to the major cities of north Italy, for a skilled clockmaker to repair the astrarium.

In whatever manner Torriano came to the Emperor's

FIG. 52. Automaton of a dancing lady with a lute, believed to be the work of Gianello Torriano, ca. 1560. Kunsthistorisches Museum, Vienna. Photograph courtesy the Kunsthistorisches Museum.

attention, it was for the purpose of restoring the astrarium. Tradition relates that Torriano assembled the scattered remnants of the masterpiece and informed the Emperor that it was beyond repair, owing to excessive corrosion. He agreed to make a similar clock, but this was not accomplished before the Emperor's departure from Italy, nor was it completed during the next several years.

It is fairly certain that Torriano did not join the Emperor's service at the time of the Emperor's visit to Italy in 1529–1530, and probably not before 1549. There exists an Imperial decree of Charles V from Innsbruck dated 7 March, 1552, assigning to Torriano annual payment of 100 gold scudi for his work.

Following his abdication, the Emperor in November, 1555, caused all his clocks and automata to be packed, presented Torriano with the sum of 1,200 gold scudi, and gave him leave of absence to visit his native city of Cremona with orders to rejoin him in Spain via Genoa.

During the sojourn of Charles V at San Yuste, Torriano devoted himself to averting the Emperor's moods of depression by creating little automata for his diversion. Tradition relates that Torriano's little figures often appeared on the dinner table after the Emperor's meal in the form of armed soldiers which marched about, rode horseback, beat drums, blew trumpets, and engaged in battle with lances. At another time Torriano is said to have released little birds carved of wood which flew about the room, out of the windows and returned, to the great disapproval of the Father Superior, who considered them to be works of the devil.

An automaton of a lady which dances and plays a lute, in the collection of the Kunsthistorisches Museum in Vienna, is believed to have been the creation of Torriano. If so, it is the only example of his automata which has survived.

When Charles V died in 1558, Torriano continued in the service of his son, Filippo II, and accompanied him to Toledo. He remained in his employment until Torriano's death, in about 1585. Torriano distinguished himself in the service of Filippo. One of his projects was the construction of a hydraulic works which furnished water from the river Tagus to the cities of Alcazar province. This is described in an article entitled "El Artificio de Juanelo y el puente de Julio Cesar" by Luis de la Escosura y Morrogh which appeared in Madrid in 1888 [91].

A general account of Torriano's life and work, compiled from secondary sources, forms part of Enrico Morpurgo's *Dizionario degli orologiai italiani* published in Rome in 1950 [55].

The astronomical clock which Torriano produced for the Emperor, inspired by the astrarium of de' Dondi, was described in 1575 by one of Torriano's contemporaries, Ambrosio de Morales [54], a most important work which has been repeatedly cited by later writers (see above, p. 40). We have recently been informed by Dr. Ladislao Reti of São Paolo, Brazil, that he has discovered unpublished manuscripts in Spain which state that Torriano's astronomical clock inspired by the astrarium had been preserved in Torriano's own home until at least the early seventeenth century. At the time of his death in 1585 Torriano had been engaged in converting the clock to the new calendar. Dr. Reti is presently working with Dr. A. G. Keller of the University of Leicester in the preparation for publication of Torriano's manuscript on hydraulic engineering (see Reti, "Postscript" [112b], p. 433 and *passim*).

The clocks constructed by Torriano for the Emperor and his son, as described in old royal inventories, are

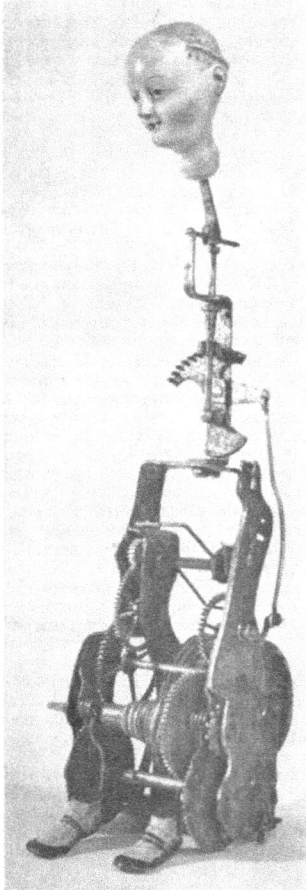

FIG. 53. Mechanism of the automaton of a dancing lady with a lute. Kunsthistorisches Museum, Vienna. Photograph courtesy the Kunsthistorisches Museum.

discussed in a book by Paulina Junquera entitled *Relojería palatina* published in Madrid in 1956 [40].

More recently, Luís Montanés Fonteñila reviewed the subject of Torriano's horological work in a comprehensive article entitled "Los Relojes del Emperador. Replanteo provisional de este tema" which appeared in Madrid in 1959 [93].

Portraits of the mechanician form the subject of an article by José Christóbal Sánchez Mayendis, "Los Retratos de Juanelo," published in 1957 [113a].

Valuable new material about Torriano's last years was recently discovered in contemporary family documents. These are quoted and described by Casto Maria del Rivero in an article entitled "Nuevos documentos de Juanelo Turriano" published in Madrid in 1936 [113].

The most recent discussion of Torriano's life and work will be found in a work by H. von Bertele and E. Neumann on an elaborate clock, supposedly made by Torriano

for the Emperor Charles V, and now in the Joseph Fremersdorf Collection at Lucerne [76a].

Tritio, Antonio de, identity not certain, but believed to have been the representative of Duke Francesco Sforza at the court of the French King in Paris in 1456.

Trizio, Paolo, fourteenth(?)-century writer, stated by Magenta [48], I, p. 569, to have written a treatise or description of the astrarium soon after Giovanni de' Dondi did so. Magenta's source is not given, and it has not proved possible to identify Paolo Trizio. However, a treatise on the astrolabe is attributed to a Paulus Tritius (see above p. 28 and note 88).

Trotto, Giacomo, of Ferrara. He was a spectator at, and possibly a participant in, the masque 'Il Paradiso' at the Castello a Porta Giovia in Milan on 13 January 1490. He is believed to have been the author of a manuscript account of the festival.

Trovamala, Don Bartholomeo, messenger or representative of Duke Francesco Sforza in 1456.

Varese, Ambrogio, da Rosate (1437–1515) was an eminent physician, philosopher and astronomer of his time. He was the author of a work entitled *Monumenta philosophiae et astronomiae.*

Ambrogio was born in Milan in 1437, the son of Bartolomeo da Rosate, physician and decurion of the city. Ambrogio excelled in his studies and was nominated lector at the University of Pavia in 1461. He was subsequently one of the four astrologers established as professors in the University by Duke Lodovico il Moro. He married Francesca Omati and produced a large family. His brother, Francesco, was also a brilliant scholar and was sent as public speaker (orator) by Duke Lodovico to the court of his father-in-law, the King of Naples.

Ambrogio Varese became the personal physician and astrologer of Duke Gian Galeazzo and later continued in the same capacity with Duke Lodovico il Moro. Lodovico's life was governed by his horoscope, and consequently by his astrologer, Ambrogio da Rosate. On 11 November, 1943, the Duke awarded to Ambrogio Varese the prefecture of Cortesella in Parma and the fief or domain of Rosate, with the privilege of imposing taxes and related benefits from the barreling of wine, the grain harvests, and of the butchering of livestock, as well as other favors and donations. In addition to the property he acquired from Lodovico il Moro, Ambrogio maintained a home in Pavia, mentioned in surviving correspondence in the Archivio di Stato of Milan. He was eventually ennobled for his services with the Duke of Milan. Ambrogio appears to have engendered the enmity of Leonardo da Vinci, as evidenced in a notation on folio 4 recto of *Manuscript B* (Institut de France). Leonardo was extremely contemptuous of the medical practices and practitioners of his time.

Of particular importance to this study is the fact that following the death of Gian Galeazzo Sforza on 22 October, 1494, the astrarium was discovered in Ambrogio's castle at Rosate, from which it was presumably removed and returned to the castle at Pavia. Whether the astrarium was acquired as a gift from the Duke Gian Galeazzo or from his wife cannot be ascertained. The letter from Lodovico il Moro (Document XXIII) leaves no doubt that he had been unaware of Ambrogio's possession of the masterpiece and that he wished to acquire it for his own (see p. 33 above).

Varese, Giovannni, da Rosate, son of Ambrogio. Da Vinci noted on the cover of *MS L* that following the overthrow of Lodovico il Moro by the French in 1500, "Giovanni da Rosate [was] robbed of all his money."

Villard dé Honnecourt, a thirteenth-century French architect and engineer, born at Honnecourt near Cambrai, and named Villard or Wilars. He collaborated in the construction and decoration of the cathedral of Rheims. His album (*ca.* 1235) of designs and sketches included a drawing of a clock housing of great interest.

Vinci, Leonardo da (1452–1519), Italian painter, sculptor, architect, engineer, and scientist. He was born in Vinci and went to Florence in 1466, where he became the protégé of Lorenzo the Magnificent. He went to Milan as the protégé of Lodovico il Moro in 1482 and continued to work for him at Milan and Vigevano until his patron's overthrow in 1499.

Wallingford. See *Richard of Wallingford.*

Wissekerke, Willelmus Aegiddii de, of Carpentras. See *Gilliszoon, Willem* and *Zelandino, Guglielmo.*

Zanino, master clockmaker employed by Cicco Simonetta on behalf of Duke Gian Galeazzo Sforza to construct a tower clock for the Castello a Porta Giovia in Milan in 1457, lived in Pavia *ca.* 1478, when it is claimed he was assigned to repair the astrarium, but did not accomplish it.

Zelandino, Guglielmo, master clockmaker claimed to have been summoned from the French court to Pavia to repair the astrarium in *ca.* 1437 and 1456. May probably be identified with Willem Gilliszoon of Carpentras (see above, pp. 25, 55).

III. CHRONOLOGICAL LIST OF EVENTS AND DOCUMENTS

ca. 1348. Giovanni de' Dondi initiated construction of astrarium at Padua.

1364. Astrarium was completed by de' Dondi at Padua.

1360–1365. Castello Visconteo constructed in Pavia for Gian Galeazzo Visconti.

1372. De' Dondi lectured at University of Pavia.

1378–1379. De' Dondi lived in Pavia for one year while attending Duke's son.

1381. Astrarium acquired from de' Dondi by Duke Gian Galeazzo Visconti.

1385. Astrarium described by Philippe de Maisieres.

1388. 11 July. Astrarium described in letter from Giovanni Manzini.

1399. Record of salary paid to Giovanni de Clarii, custodian of astrarium.

1420. Pier Candido Decembrio noted presence of the astrarium in the ducal library at Pavia.

1440. Michele Savonarola noted that several years earlier the astrarium was in disrepair and had been restored to operating condition by a clockmaker from France.

1456. 17 April. Facino da Fabriano, chancellor, communicated to Duke Francesco Sforza regarding repairs needed in library, including installation of an arrangement for lifting the cover of the astrarium on and off.

1456. 29 April. Duke wrote to Facino to send him the book from ducal library relating to "our clock."

1456. 1 May. Da Fabriano communicated with Duke concerning the hiring of a suitable repairer for the astrarium, and regarding the damage already done in the library by the cover of the astrarium.

1456. 12 May. Letter from Duke's secretary to French Court requesting that a clockmaker whom he called Guglielmo of Paris be sent to Pavia to repair the astrarium.

1459. 28 May. Cicco Simonetta wrote to Conte Bolognino de Attendolis to find and hire a custodian for the astrarium and another to maintain the collections of books.

1460. 6 January. Letter to Conte Bolognino de Attendolis to show the clock in the library and other reliquaries to Master Marco de Raynis of Milan.

1463. 26 September. Cicco Simonetta on behalf of Duke, wrote to an abbot in Brescia to request return of a book probably on de' Dondi clock.

1463. Astrarium inspected by Johann Müller called Regiomontanus and subsequently referred to in his introductory lecture at Padua.

1464. 5 November. Francesco Sforza wrote to Conte Bolognino to give tutor of his children various books from library, including "the book about the clock."

1467. Name of Gaspare d'Allemagna *custos relogii* (clock custodian) listed among employees in the service of Duke Galeazzo Maria Sforza.

1471. 14 February. Duke wrote to Francesco Sforza approving request to borrow book of the "Astrolabio" from the library in Pavia.

1474. Regiomontanus noted in a publication the construction then in progress of his own astronomical clock inspired by the astrarium in Nuremberg.

1476. The Regiomontanus clock remained unfinished at the time of his death.

1489–1490. Leonardo da Vinci worked in ducal library in Pavia and probably saw the astrarium.

1490. Leonardo da Vinci designed mechanism and stage setting for masque "Il Paradiso" possibly based upon astrarium, or manuscript volume describing it which was in ducal library.

1489–1490 or 1494–1495. Da Vinci may have made sketches of the dials of Venus and Mars of the de' Dondi astrarium.

1492. Reconstruction of castle at Vigevano was initiated and Leonardo and Bramante were employed by Duke for this project.

1493. Bernardino Bellincioni wrote his *Rime* which described "Il Paradiso" and Leonardo's role in designing it.

1494. October. Duke Gian Galeazzo Sforza died suddenly at Pavia, and was succeeded by his uncle, Lodovico il Moro.

1494. 6 November. Duke Lodovico il Moro advised Gualtiero Bescapé by letter that he had just learned that the de' Dondi clock was in a hall at Rosate, and that he would decide on his course of action upon his return to Milan.

1495. 4 March. Bianchino da Palude at Vigevano reported a trip made by Bramante to Pavia to obtain several items from son of Ambrogio Varese of Rosate.

1495. 5 March. Jacopo da Pusterla, custodian, reported visit made by Bramante to ducal library to sketch astrarium, and his refusal to permit sketches to be removed.

1495. 6 March. Duke advised da Pusterla to permit Bramante to make and take whatever sketches of the astrarium he wished.

1529 or 1530. Emperor Charles V in Italy for his coronation at Bologna, saw the astrarium. He requested a clockmaker to restore it. Gianello Torriano of Cremona, who responded to the request, reported that it could not be repaired, and subsequently produced a reconstruction based on the original.

IV. THE RULERS OF MILAN

1322–1328. *Galeazzo Visconti I* (1277–1328), was the successor of his father, Matteo Visconti (1250–1322). He married Beatrice d'Este, widow of Nino di Gallura referred to in the "Purgatorio" of Dante's *Divina Commedia.* He was subsequently succeeded by his son.

1328–1339. *Azzo Visconti* (1302–1339). Azzo purchased the city and the title of imperial vicar from Emperor Louis the Bavarian, and conquered ten Lombard towns during his reign. In 1329 he murdered his uncle, General Marco Visconti, and suppressed a revolt led by his cousin, Lodrisio. He built the tower of San Gottardo in Milan.

After his death, his two uncles, Lucchino and Giovanni, divided the rule of Milan between them.

1339–1349. *Lucchino Visconti* (1287–1349) became temporal ruler of Milan; his brother Giovanni (1290–1354), who was Archbishop of Milan, became the spiritual ruler. Lucchino married Isabella Fieschi and resided in S. Giovanni in Conca. He succeeded in adding Parma and Pisa as dependencies of Milan. While he was planning the murder of his unfaithful and disloyal wife, she succeeded in poisoning him. Since Lucchino had sons but not of proven legitimacy, the rule of Milan went to his brother,

1349–1354. *Giovanni Visconti* (1290–1354), one of the most notable characters of the fourteenth century. He lived at the Arcivescovado where he maintained a hospitable court, and where his friend, Francesco Petrarch, was a prominent figure. Giovanni ascended the archepiscopal throne in the Arcivescovado with a crosier as his spiritual sceptre in one hand and a drawn sword as his temporal sceptre in the other. By extending Visconti rule to Bologna in 1350, defying Pope Clement VI, and annexing Genoa in 1353, Giovanni established the Visconti family as sovereigns of the whole of northern Italy except for Venice, Piedmont, Verona, Mantua, and Ferrara. Upon his death his rule was divided between the three sons of his deceased brother, Stefano. Matteo II, the eldest, became ruler of Bologna, Lodi, Piacenza and Parma. After engaging in a career of bestial sensuality and immorality, he was assassinated in 1355 by the order of his brothers. Bernabò ruled Cremona, Crema, Brescia and Bergamo, while Galeazzo held Como, Novara, Vercelli, Asti, Tortona, and Alessandria. After Matteo's death, they continued to rule jointly and capably.

1354–1378. *Galeazzo Visconti II* (1320–1378), who was considered to be the handsomest man of his age, resided at his castle at Pavia. He is remembered in history as the patron of Petrarch and as the founder of the University of Pavia. He married Bianca of Savoy. His daughter, Violante, was subsequently married to the Duke of Clarence, son of King Edward III of England. His son, Gian Galeazzo, married Isabella, daughter of King John of France. Galeazzo died in 1378, and the Visconti rule of Milan continued jointly under his son and the third brother.

1354–1385. *Bernabò Visconti* (1323–1385) displayed the worst vices of the Visconti and was noted for his cold blooded cruelty. He maintained his court at the Castello a Porta Giovia in Milan. His career was marked with constant warfare and strife, resulting in oppressive taxation of his people. He fought against Popes Innocent VI and Urban V, who declared a crusade against him; and against Emperor Charles IV, who took away Bernabò's fief. Following the death of Galeazzo II, Bernabò attempted to usurp the rule from his nephew, Gian Galeazzo, who lived in Pavia during the joint rule and had made countless friends with his kindness and humanity. After the young ruler succeeded in making his uncle a prisoner, the latter's palace was sacked and the members of his family killed by the people. Bernabò died seven months later in prison.

1378–1385. *Gian Galeazzo Visconti,* called Conte di Virtù (1347–1402), was a sedate but crafty ruler with a great love of order and precision. He was an intellectual, and he undertook great engineering projects which remain as monuments to his memory to the present time. He was a master of intrigue, yet exhibited great personal timidity. He ruled Milan jointly with his uncle, Bernabò, following his father's death. He was married first to Isabella of France and then to Caterina Visconti. Following the death of Bernabò in 1385,

1385–1402. *Gian Galeazzo* ruled alone. His reign marked the greatest period of power of the Visconti family. In 1395 he received the title of Duke of Milan from Emperor Wenceslaus for the price of 100,000 florins.

The death of his three sons by Isabella left him no male heirs and as a result he undertook the construction of the Cathedral of Milan in 1384. Though the first stones were laid in 1386, the building was not completed until 1927. In 1395 he built the Certosa at Pavia as well as the bridge over the Ticino River. He enlarged and improved the University of Pavia, established a library there and restored the University of Piacenza. His great ambition was to subdue all of Italy under his family rule, and he succeeded in conquering many of the important cities of northern Italy. He conquered Verona in 1387, and took Padua with Venetian aid in 1388. After subduing Mantua and Ferrara, he bought Pisa, conquered Siena, and in 1400–1401 seized Perugia, Lucca and Bologna. He was besieging Florence when he died of the plague in 1402. He was succeeded by

1402–1412. *Giovanni Maria Visconti* (1389–1412), his older son, who was a minor at the time of his father's death. Until he was of age, in 1407, Giovanni Maria reigned under the regency of his mother, Caterina Visconti, and the protection of the condottiere, Facino Cane de Cesale. As the second Duke of Milan, Giovanni Maria exhibited extreme sadism and particularly enjoyed seeing criminals being torn to pieces by savage wolfhounds. Tradition relates that he roamed the streets at night with a huntsman and a pack of savage dogs which he unleashed on anyone moving about. At the age of twenty-four he was assasinated by three Ghibelline partisans and his body was flung into the cathedral constructed by his father. In 1412 the duchy went to his younger brother.

1412–1447. *Filippo Maria Visconti* (1391–1447), who had become the nominal ruler of Pavia upon his father's death, under the protection of the condottiere, Facino Cane. He was extremely ugly and very sensitive about his appearance. His reign is marked by his insane cruelty and infernal lust. At the same time he was brilliant, crafty and a good judge of men. Although he feared them, he employed the most capable generals. He maintained a successful system of secret intelligence and had a number of astrologers and sorcerers to counsel him.

He lived and worked in secret and was seldom seen by the public, for he feared everyone. In spite of these characteristics, Filippo was an astute politician and succeeded in recovering the Lombard cities which his father had conquered and which were lost again upon his death. Filippo Maria lived in the Castello a Porta Giovia in Milan, a fortified citadel which was the center of intrigues. From his marriage to Beatrice di Tenda, the widow of Facino Cane, he acquired a considerable dowry. He had Beatrice beheaded in 1418 on the basis of trumped up charges and then married Agnese del Maino. She had been his mistress for many years and the mother of his illegitimate daughter, Bianca Maria, who later married Francesco Sforza. Filippo Maria is remembered as a serious scholar of ancient authors, and for his patronage of local silk manufacture and other industries. He was the last in the male line of the Visconti, and when he died in 1447 from an infected wound, Milan was ruled by

1447–1450. The Ambrosian Republic, established by the people. In 1450 the Republic was overthrown by

1450–1466. *Francesco Sforza* (1401–1466), a former condottiere, who became the fourth Duke of Milan. Previously he had fought against the Venetians under Filippo Maria Visconti (and against the Visconti for the Venetians). He was the ablest general of his time, a firm and wise ruler and a master of statecraft. In 1441 he married

Bianca, the only daughter, although illegitimate, of Filippo Maria Visconti, and received Cremona and Pontremoli as a dowry. She was seventeen and he was forty when they married; and he was already the acknowledged father of twenty-two bastard children. Following the death of Filippo Maria, the Milanese had proclaimed a republic and torn down the Visconti castle. After three unsettled years of democratic rule the city was on the verge of starvation. Fearing invasion from their enemies and neighbors, they were glad to relinquish the city to Francesco Sforza, who enjoyed the friendship and financial support of Cosimo de' Medici of Florence. Duke Francesco ruled frugally, cast aside the astrologers, and won over the people with his affability and accessability. He employed three thousand men to rebuild the castle, while he lived with Bianca at a small palace called the Corte d'Arego. His court was filled with Greek exiles, important scholars, and the foremost intellectuals of the time, including Giovanni Averulino (Filarete), Guiniforte Solari and such well-known humanists as Pier Candido Decembrio and Francesco Filelfo. The latter served as tutor to the Duke's numerous children. Francesco died in 1466 at the age of sixty-five, following a severe attack of dropsy. He left several sons, of whom the eldest and his successor was

1466–1476. *Galeazzo Maria Sforza* (1444–1476), who was in France at the time of his father's death. Summoned by his mother, he returned to Milan in disguise, wearing mourning and riding a black horse. He was met at the gates of the city with the ducal emblems, whereupon he changed into the garments of his new office and rode into the city on a white charger to become the fifth Duke of Milan. He was twenty-two years old.

Galeazzo Maria married Bona of Savoy, the sister-in-law of King Louis XI of France, at the latter's suggestion. The young Duke had not met her but she was reported to be of great beauty according to the Milanese ambassador who noted in his report, however, that he had seen her only from the front. Galeazzo's brother, Tristano, went to France to marry Bona by proxy. Upon her arrival in Milan, the castle was still under construction, and the young couple spent their honeymoon on an island in the park. Galeazzo Maria inherited none of his father's personality, and consequent popularity, but resembled his Visconti forbears. He built his court into the most splendid in Europe. He was a great lover of the arts and established a school of music to which singers flocked from afar. He encouraged graphic arts, and under his sponsorship such notable printers as Panfilo Castaldi and the Zaroto brothers came to Milan to work. Galeazzo Maria was equally noted for his eloquence, extravagant lust and vile cruelty. His reign ended after a decade with his assassination while attending High Mass in the Church of San Stephano. When he entered the church, a man stepped forward and knelt before him. As the Duke paused, the knives of three assailants were buried in his body and he died at once. The murderers were seized, hanged, drawn and quartered. The instigator of the murder proved to be a humanist named Cola Montana, who nurtured a grievance against the Duke. He had incited three of his students to re-enact the classic assassination of Caesar with the Duke as the victim. The Duke's illegitimate daughter, Caterina, Countess of Sforza, married Girolamo Riario, son of Pope Sixtus IV, and subsequently played an important role in the history of fifteenth-century Italy. The Duke's murder marked the decline of Italy's prosperity and paved the way for subsequent foreign rule. The Duke was succeeded by his seven-year-old son,

1476–1494. *Gian Galeazzo Sforza* (1468–1494), under the regency of his mother, Bona of Savoy. She was noted for her beauty and gaiety but was not distinguished for her good sense. She became enamored of a handsome young servant in her household named Tassino, to whom she gave gifts and privileges to such an extent that the nobles of the court forced him to leave the city. He took with him a fortune in gems. Bona tried to follow him and spent the remainder of her life in obscurity and disillusion. As a result of this situation, the guardianship of the young Duke was assumed by the boy's uncle, Lodovico il Moro. In 1489 Gian Galeazzo married Isabella of Aragon, the daughter of King Alfonso of Calabria and of Naples. They resided in the castle at Pavia, and the young Duke devoted himself to the pursuit of pleasure and left the rule of the duchy to his uncle. Upon his death in 1494 the duchy of Milan went to

1480–1494. *Lodovico Sforza* (1452–1508), called 'il Moro' or 'the Moor.' He had been named Lodovico Mauro Sforza and as a pun on his second name, Lodovico took as his emblems a Moor's head and a mulberry tree (*gelsa mora*), and he used Moorish or African servants. After reigning jointly with Gian Galeazzo for fourteen years, he reigned alone from

1494–1499. In 1491 Lodovico had married the sixteen-year-old Beatrice d'Este, younger daughter of the Duke of Ferrara. It was a happy marriage and the couple resided first at Vigevano and later at Pavia. The duchy was wealthy during this period, and Lodovico sponsored numerous important engineering and architectural enterprises throughout the duchy. He became the patron of Leonardo da Vinci, who worked under his sponsorship in Milan for sixteen years; and of Donato Bramante. Before Beatrice's death in 1497 Lodovico was the father of two sons, Massimiliano and Francesco Maria. Meanwhile the King of Naples feared that Lodovico would usurp the duchy from his sickly nephew and threatened war to protect the young Duke and his Neapolitan bride. As a result, Lodovico invited King Charles VIII of France to come to Italy to reassert the French claim to Naples. The arrival of the French coincided with the death of Duke Gian Galeazzo in 1494. Although Lodovico offered his loyalty to the Duke's infant son, Milan's Council of Ten asked him to become Duke instead. Meanwhile the French army captured Naples, and Lodovico, finding that the French policies endangered his own position, allied against Charles VIII. He gave his niece, Bianca, in marriage to Emperor Maximillian I and obtained imperial investiture of the duchy. In 1497, the year after the French left Italy, Beatrice died in childbirth. The following year Charles VIII died and was succeeded by Lodovico's enemy, the Duke of Orleans, who became King Louis XII. Surrounded by his enemies, Lodovico fled Milan. He was reinstated briefly by the Swiss and eventually delivered over by them to the French in April 1500. He died a prisoner in the castle of Loches in 1508. After his capture

1499–1512. Milan was ruled by the Swiss, until

1512–1515. *Massimiliano Sforza* (1493–1530), was restored to the rule of Milan by the Swiss. After the overwhelming defeat of his allies at Marignano in 1515 he surrendered his rights to Francis I. He died in Paris in 1530.

1515–1521. Milan was under the French rule until

1521–1535. *Francesco Maria Sforza II* (1495–1535), Lodovico's second son, was invested as ruler of Milan by the French after their defeat at La Bicocca in 1522. He reigned until his death in 1535, which marked the end of the Sforza male line, and the duchy went to Emperor Charles V of Spain.

V. CHRONOLOGY OF CLOCKWORK

Following is a chronological listing of important horological dates which can be associated with the history of the astrarium of de' Dondi. It is limited to those horological events which appear to be significant in relation to the astrarium.

ca. 65 B.C. Greece. Fragments of a geared planetary device were recovered near the Greek island of Antikythera which are believed to be Hellenistic in origin and dating *ca.* 65 B.C.

ca. A.D. 1000. Islam. al-Birûni devised calendrical gearing to show the revolutions of the sun and moon at their relative rates and to show the changing phases of the moon.

1221–1222. Işfahân. Muḥammed b. Abi Bakr of Işfahân made a brass astrolabe with a geared calendar movement based upon the design of al-Birûni.

ca. 1235. France. Drawings in the album of Villard (or Wilars) of Honnecourt near Cambrai illustrated a Gothic clock housing, and the earliest known attempt to produce controlled reciprocal motion.

ca. 1250. France. A miniature painting in a moralized Bible from northern France illustrates a clepsydra with wheelwork and a striking mechanism.

1271. France. In his commentary on the *Tractatus de Sphera* of Sacrobosco, Robertus Anglicus indicated that the mechanical clock had not yet been perfected, although attempts to do so were then in progress. Whether Anglicus was concerned with a truly mechanical escapement or with a cylindrical clepsydra has not been ascertained. In any case, it may be concluded that the mechanical escapement was not known to him, therefore providing a *terminus ante quem non* for that invention.

1276–1277. Spain. The *Libros del saber* of King Alfonso X of Castile described and illustrated an astrolabic clock regulated by a compartmented cylindrical clepsydra containing mercury.

Before 1300? North Italy. Geared planetarium, ostensibly also a time-piece with the strong possibility of an astrolabe dial in the anaphoric clock tradition.

ca. 1300. France. An unknown maker produced an astrolabe, now in the Science Museum, London, in which the relative motions of the sun and moon are indicated by pointers and which are moved by an arrangement of gears.

1309. Milan. An iron clock was installed in the campanile of the church of Sant'Eustorgio in Milan. This date is sometimes given as 1306.

1317–1320. Italy. Dante Alighieri referred to clocks and to a striking mechanism in the "Paradiso" of his *Divina Commedia* (Canto X, line 139, and Canto XXIV, line 13).

ca. 1327–1330. St. Albans. Richard of Wallingford constructed an astronomical clock for the Abbey of St. Albans in England.

1327. Oxford A record of Merton College, Oxford, mentioned the removal of the "horologium" from the Hall. This could be a reference to a work of Richard of Wallingford.

1335. Milan. A clock was installed in the campanile of San Gottardo in Milan for Azzone Visconti, apparently by Guglielmo Zelandino.

1335. Milan. A striking clock was constructed for the chapel of the Visconti palace in Milan, which indicated and struck the twenty-four hours.

1343. Modena. A clock was installed in the Duomo at Modena by Giovanni degli Organi, engineer of the city of Milan.

1344. Padua. Jacopo de' Dondi completed the design of a complicated tower clock which in March of this year is claimed to have been constructed and installed over the the gateway of the Palazzo Capitanato in Padua by a young local artisan named Antonio.

1345. Orvieto. A clock was constructed for the Torre Maurizio.

1347. Monza. Giovanni, master clockmaker of the Visconti family of Milan, constructed the clock for the Torre della Piazza del Duomo in Monza. This clockmaker may have been Giovanni da San Vincenzo.

ca. 1348. Padua. Giovanni de' Dondi began the construction of the astrarium, which was completed in 1364.

1349–1396. St. Albans. An entry in the *Gesta Abbatum* of the Abbey of St. Albans stated that the upper dial and "the wheel of fortune" of the astronomical clock of Richard of Wallingford were completed after his death.

1350–1400. Pavia. Andrea degli Organi, son of Giovanni, made repairs to the clock of the Visconti castle at Pavia.

1353. Milan. A clock was installed in the Torre della Piazza at Milan.

1354. Genoa. Giovanni degli Organi constructed a public clock for the tower of the Duomo of San Lorenzo in Genoa when that city fell under the protectorate of the Visconti.

1354–1357. Genoa. Ambrogio, master clockmaker of Milan, was the first custodian of the clock in the Duomo at Genoa.

1354. Florence. Niccolo Bernardi (Berardi?) of Florence constructed a clock for the tower of the Palazzo Vecchio at Florence.

1355–1371. Reggio Emilia. The first public clock in Reggio Emilia was constructed.

1356. Perpignan. A clock and bell were installed in the castle of Perpignan by order of Pedro (Pere) III, *el ceremonioso*. The construction of the clock was entrusted to Antonio Bovelli of Avignon, *plomberius* to Pope Innocent VI.

1356. Bologna. According to the *Chronica miscella bononiensis*, the great bell of the tower of the Palazzo Biada was removed and installed on the Torre dell' Capitano as part of the first public clock of the city, which was being erected at this time.

1357. Genoa. Gregorio Durnasio was appointed moderator of the clock of San Lorenzo in Genoa, and was succeeded in this position in 1364 by Giovanni di San Vincenzo.

1358. Florence. Giovanni di Pacino of Milan repaired the tower clock of the Palazzo Vecchio in Florence.

1359. Siena. Bartolo Giordi (or Giudi) reputedly constructed a clock for the municipal tower at Siena.

1360. Siena. The public clock on the Torre della Manzia at Siena, was reputed to have been the first public clock of that city, and to have been constructed by one Perino in 1360.

1362. Ferrara. The clock in the Torre del Palazzo d'Este, which had been installed at an earlier unknown date, was reported not to be operating properly.

1362–1370. Paris. Heinrich von Wiek (Henry de Wick) of Germany constructed a clock for the palace of King Charles V, now the Palais de la Justice, in Paris.

1363. Siena. A monk named Luca dello Spedale was appointed custodian of the clock on the Torre della Manzia at Siena and served until 1369 when he was succeeded by the jeweller and mosaicist, Michele de Ser Memmo, who served until 1376.

1364. Padua. Giovanni de' Dondi completed the construction of his astrarium.

1368. London. King Edward III invited three clockmakers of Delft to work in England and provided them with a patent for safe conduct.

1369. France. In the poem "Li Orloge Amoureus," Jean Froissart compared the various parts of a clock to the attributes of love. The clock he described had the elements of a weight-driven clock regulated by a foliot, and striking work of bells struck by hammers operated by pins.

1369. Udine. A public clock was installed on the Torre del Orologio in Udine.

ca. 1370. Strassburg. An astronomical clock was erected in the Cathedral of Strassburg.

1381. Pavia. Duke Gian Galeazzo Visconti acquired the astrarium of Giovanni de' Dondi and installed it in the library of the castle at Pavia.

1399. Pavia. Joannes de Clarii was custodian of the tower clock in the Visconti castle at Pavia, and perhaps of the astrarium as well.

1423. Padua. A design for a striking clock submitted by Novello dall'Orologio was accepted by the city council of Padua to replace the clock of Jacopo de' Dondi, which was destroyed in 1390.

1430–1434. Padua. The construction and installation of the clock designed by Novello dall'Orologio was assigned to Giovanni dalle Caldaje in 1430 who completed the work in four years.

1457. Pavia. Zanino, a master clockmaker, was employed by Cicco Simonetta to construct a tower clock for the Castello a Porta Giovia at Milan for Duke Gian Galeazzo Sforza.

1467. Pavia. Gaspare d'Allemagna was a monk listed as "custodian of the clocks" on the rolls of Duke Galeazzo Maria Sforza at his castle at Pavia. By 1470 he was no longer listed.

1475. Milan. Claudio, a master clockmaker in Milan, was banished from the Dukedom.

late 1400's. Bologna. A monastic alarm clock of the late fifteenth century, similar in design and construction to a timepiece depicted in intarsia panels executed between 1528 and 1543 by Fra Damiano Frambelli of Bergamo in the Basilica of San Domenico in Bologna, was produced.

1480–1484. Florence. Lorenzo della Volpaia completed the construction of a complicated astronomical clock showing the motions of the planets, which was subsequently installed in the Palazzo Vecchio.

VI. SOURCES AND BIBLIOGRAPHY

BOOKS

1. AMATI, G. 1828. *Ricerche storico-critico-scientifiche sulle origini, scoperte, invenzioni, ecc.* (Milan).

2. AMEISENOWA, ZOFIA. 1959. *The Globe of Martin Bylica of Olkusz and Celestial Maps in the East and West.* (Polska Akademia Nauk. Komitet Historii Nauki. Monografie z dziejów nauki i techniki XI), Wrocław, Cracow & Warsaw.

3. BARNI, LUIGI. 1949. *Piazzi e torre di Vigevano* (Vigevano, Francini).

4. —— 1951. *Vigesimum* (Vigevano, Francini).

5. BAILLIE, GRANVILLE HUGH. 1951. *Clocks and Watches, An Historical Bibliography* (London, N.A.G. Press, Ltd.).

6. ——. 1929. *Watches, Their History, Decoration and Mechanism* (London, Methuen & Co.).

7. BALDELLI, S. 1837. *Vita di Francesco Petrarca* (Fiesole).

8. BARUCCI, GALILEO. 1909. *Il Castello di Vigevano nella storia e nell'arte* (Turin).

8a. BARYCZ, HENRYK. 1933. *Conclusiones Universitatis Cracoviensis ab anno 1441 ad annum 1589* (Polska Akademja Umiejetnosci. Archivum Komisji do Dziejów Oswiaty i Szkolnictwa w Polsce No. 2, Cracow).

9. ——. 1935. *Historja Universytetu Jagiellońskiego w epoce humanizmu* (Cracow).

10. —— 1957. *The Development of University Education in Poland* (Warsaw).

11. BARZON, ANTONIO, ENRICO MORPURGO, ARMANDO PETRUCCI and GIUSEPPE FRANCESCATO. 1960. *Giovanni Dondi dall' Orologio, Tractatus astrarii . . .* Biblioteca Capitolare di Padova, Cod. D. 39. *Introduzioni, trascrizione e glossario a cura di Antonio Barzon, Enrico Morpurgo, Armando Petrucci, Giuseppe Francescato, con la riproduzione fotografica del codice* (Codices ex ecclesiasticis Italiae biblithecis selecti, phototypice expressi . . . vol. IX, Vatican City).

12. BASSERMAN-JORDAN, ERNST. 1905. *Die Geschichte der Räderuhr unter besonderer Berücksichtigung der Uhren des Bayerischen Nationalmuseums* (Frankfurt am Main).

12a. BEAUJOUAN, GUY. 1962. *Manuscrits scientifiques médievaux de l'Université de Salamanque et de ses "Colegios mayores."* Bibliothèque de l'École des Hautes Études hispaniques 32 (Bordeaux).

13. BECKMANN, JOHN. 1846. *A History of Inventions, Discoveries, and Origins* (2 v., 4th ed., London, Henry G. Bohn).

14. BELLEMO, VINCENZO. 1894. *Jacopo e Giovanni de Dondi dall' Orologio. Note critiche* (Chioggia).

15. BELLINCIONI, BERNARDINO. 1493. *Sonetti, Canzoni, Capitoli* (Milan).

16. —— 1878. *Le Rime* (Bologna).

17. BELTRAMI, LUCA. 1894. *Il Castello di Milano durante il dominio dei Visconti e degli Sforza* (Milan, Hoepli).

18. BOWIE, THEODORE, ed. 1962. *The Sketchbook of Villard de Honnecourt* (2nd ed., New York).

19. BREVENTANO, STEFANO. 1570. *Istoria delle antichità, nobilità, e delle cose notabili della città di Pavia* (Pavia, Hieronimo Bartholi).

20. BUGATTI, G. 1571. *Istoria universale* (Venice).

21. CARNERALLI, VICENZO. 1932. *La Sforzesca* (Vigevano).

22. CARDANUS, HIERONYMUS. 1554. *De Subtilitate* (Lyons).

23. CASATI, C. 1870. *I Capi d'arte di Bramante da Urbino nel Milanese* (Milan, Hoepli).

24. CICOGNARA, CONTE LEOPOLDO. 1823. *Storia della scultura dal suo risorgimento* (Prato, Giachetti).

24a. CLARK, SIR KENNETH. 1935. *A Catalogue of the Drawings of Leonardo da Vinci in the Collection of His Majesty the King at Windsor* (Cambridge, Cambridge University Press).

25. CLAUSSE, GUSTAVE. 1909. *Les Sforza et les arts en Milanais 1450–1530* (Paris, Ernst Leroux).

26. COLLE, FRANCESCO MARIA. 1824–1825. *Storia scientifico-letterario dello Studio di Padova* (Padua, 4 v. Tip. della Minerva).

27. COXE, HENRY O. 1858–1885. *Catalogi codicum manuscriptorum Bibliothecae Bodleianae, pars secunda. Codices latinos et miscellaneos Laudianos* (Oxford).

28. COSENZA, MARIO E. 1962. *Dictionary of the Italian Humanists* (New York).

29. D'ADDA, GEROLAMO. 1875. *Indagini storiche, artistiche e bibliografiche sulla libreria visconteo-sforzesca* (Milan).

30. —— 1879. *Indagini storiche, artistiche e bibliografiche sulla libreria visconteo-sforzesca* (Milan).

31. DECEMBRIO, PIER CANDIDO. 1731. *Vita Philippi Mariae Vicecomitis Mediolanensium Ducis III* in Muratori, *Rer. Ital. Script.* [56], Vol. XX, Milan.

32. DE'DONDIS, GIOVANNI. 1553. *De Balneis* (Venice, Junctae).

33. DELL'ACQUA, CARLO. 1939. *Vigevano nella storia, nell'arte e nell'industria* (Vigevano).

33a. EDWARDES, ERNEST L. 1965. *Weight-driven Chamber Clocks of the Middle Ages and Renaissance. With some observations concerning certain Larger Clocks of Medieval Times* (Altrincham).

34. ERIZZO, NICCOLÒ. 1860. *Relazione storico-critica della Torre dell'Orologio di S. Marco in Venezia* (Venice, Tip. dell'Commercio).

34a. FONTENAI, L'ABBÉ DE. 1776. *Dictionnaire des artistes, ou notice historique et raisonnée des architectes, peintres, graveurs, sculpteurs, musiciens, acteurs & danseurs; imprimeurs, horlogers & mechaniciens* (Paris, 2 vols.).

35. GABOTTO, FERDINANDO. 1891. *Nuove ricerche e documenti sull'astrologia alla corte degli Estensi e degli Sforza* (Turin).

36. GIMMA, GIACINTO. 1723. *Idea della storia dell'Italia letteraria* (Naples, F. Mosca).

37. GLORIA, ANDREA. 1885. *L'Orologio di Jacopo Dondi* (Padua).

38. —— 1888. *Monumenti della Università di Padova 1318–1405* (Padua).

38a. GUNTHER, R. T. 1923. *Early Science in Oxford* (Oxford), Vol. II.

39. JAMES, MONTAGUE RHODES. 1895. *A Descriptive Catalogue of the Manuscripts in the Library at Eton College* (Cambridge).

40. JUNQUERA, PAULINA. 1956. *Relogería palatina* (Madrid, Roberto Carbonell Blasco).

40a. KRISTELLER, PAUL OSKAR. 1963. *Iter italicum. A Finding List of Uncatalogued or Incompletely Calalogued Humanistic Manuscripts of the Renaissance in Italian and Other Libraries* (London and Leyden).

41. [LAZZARI, ABBATE, ed.] 1854. *Miscellaneorum ex manuscriptis libris bibliothecae Collegi Romani Societatis Jesu* (Rome).

42. LLOYD, H. ALAN. 1958. *Some Outstanding Clocks Over Seven Hundred Years, 1250–1950* (London, Leonard Hill).

43. —— 1964. *The Collector's Dictionary of Clocks* (London, Country Life Limited).

44. McCURDY, EDWARD. 1939. *The Notebooks of Leonardo da Vinci* (New York, Reynal and Hitchcock).

45. —— 1940. *The Mind of Leonardo da Vinci* (New York, Dodd Mead & Co.).

46. MADAN, FALCONER, and H. H. E. CRASTER. 1922. *Summary Catalogue of Western Manuscripts in the Bodleian Library*, Vol. III, Pt. 1, (Oxford).

47. MADDISON, FRANCIS R. 1957. *A Supplement to a Catalogue of Scientific Instruments, In the Collection of J. A. Billmeir, Esq., C. B. E.* (Oxford and London).

48. MAGENTA, CARLO. 1883. *I Visconti e gli Sforza nel Castello di Pavia e loro attinenze con la Certosa e la storia cittadina* (Milan, Hoepli).

49. MALAGUZZI VALERI, FRANCESCO. 1913. *La corte di Lodovico il Moro: la vita privata e l'arte a Milano nella metà del quatrocento* (Milan, Hoepli).

50. —— 1915. *La corte di Lodovico il Moro: Bramante e Leonardo da Vinci* (Milan, Hoepli).

51. MANZINI, GIOVANNI. See Lazzari, Abbate.

52. MONTUCLA, JEAN ETIENNE. 1799–1802. *Histoire des mathématiques* (Paris).

53. MOORAT, S. A. J. 1962. *Catalogue of Western Manuscripts on Medicine and Science in the Wellcome Historical Medical Library* (London).

54. MORALES, AMBROSIO DE. 1573. *Las Antigüedades de las cuidades de España* (Madrid).

55. MORPURGO, ENRICO. 1950. *Dizionario degli orologiai italiani* (Rome, La Clessidra).

56. MURATORI, LUIGI A. 1723–1751. *Rerum italicarum scriptores ab anno aerae christianae 500 ad annum 1500* (Milan), Lib. I, col. 1165.

57. NEGRI, L. 1908. *Rosate e la sua pieve* (Saronno).

57a. ORE, OYSTEIN. 1953. *Cardano, The Gambling Scholar* (Princeton, N. J., Princeton University Press).

57b. PEDRETTI, CARLO. 1957. *Leonardo da Vinci Fragments at Windsor Castle from the Codex Atlanticus* (London, Phaidon Press).

57c. PELLEGRIN, ÉLISABETH. 1955. *La Bibliothèque des Visconti et des Sforza, ducs de Milan, au XVe siècle* (Publications de l'Institut de Recherche et d'Histoire des Textes V), Paris.

58. PEROGALLI and BESCAPÉ. 1960. *Castelli della pianura Lombarda* (Milan, Ed. Electa), p. 192.

59. PIERTRUCCI, NAPOLEONE. 1858. *Biografia degli artisti padovani* (Padua).

60. PIETRAGRASSA, GIOVANNI BATTISTA. 1760. *Annotazioni diverse spettanti alla fondazione della R. città di Pavia, con alcuni accidenti accaduti alla stessa città, e narrative di alcune preclare gesta di varj personaggi, nonchè della più cospicue ed antiche famiglie, con altre storiche curiosità cavate dai più famosi e chiari autori tra gli altri il rinomatissimo Volaterrano. Opera messa insieme dall'eruditissima penna del G. C. lettore pubblico nella Regia Università della sopredetta città, il signor Gio. Batt.' Pietragrassa nell'anno 1636 ed accuratamente trascritta da Leopoldo Arena, pavese, nell'anno 1760* (Biblioteca Universitaria di Pavia).

61. [POLIZIANO, ANGELO]. 1498. *Omnia opera Angeli Politiani, et alia quaedam lectu digna* (Venice).

62. RAVAISSON-MOLLIEN, CHARLES 1890–1891. *Les Manuscrits de Leonardo da Vinci, de l'Institut de France, Mss G. L. et M.* (Paris).

62a. [REGIOMONTANUS, JOHANNES]. 1537. . . . *Rudimenta astronomica Alfragani. Item Albategnius astronomus peritissimus de motu stellarum, ex obseruantibus tum proprijs, tum Ptolemaei, omnia cum demonstrationibus Geometricis & Additionibus Ioannis de Regiomonte. Item Oratio introductoria in omnes scientias Mathematicas Ioannis de Regiomonte, Patauij habita, cum Alfraganum publice praelegeret. Eiusdem utilissima introductio in elementa Euclidis. Item Epistola Philippi Melanthonis nuncupatoria, ad Senatum Noribergensem. Omnia iam rescensprelis publicata* (Nuremberg).

62b. RILEY, HENRY THOMAS, ed. 1867 & 1869. *Chronica Monasterii S. Albani, Gesta Abbatum Monasterii Sancti Albani, a Thomas Walsingham, regnante Ricardo secundo, ejusdem ecclesiae precentore compilata* (London, Rolls Series). Vols. II & III.

62c. RIVOLTA, A. 1933. *Catalogo dei codici Pinelliani dell'Ambrosiana* (Milan).

62d. ROBERTSON, J. DRUMMOND. 1931. *The Evolution of Clockwork, with a special section on the Clocks of Japan . . . together with a Comprehensive Bibliography of Horology . . .* (London, Cassell & Company, Limited).

63. SACCHETTI, EGIDIO. 1648. *Vigevano illustrato* (Milan).

64. SACCUS, BERNARDUS. 1587. *De Italicarum rerum varietate et elegantia . . .* (Pavia).

65. SARZOSIUS, FRANCISCUS. 1526. *Franciscus Sarzosi Cellani Aragonei in aequatorem planetarum libri duo; prior fabricam aequatoris complectitur, posterior usum atque utilitatem hoc est veros motus ac passiones in zodiaci decursu contingentes aequatoris ministerio investigare docet* (Paris).

66. SAVONAROLA, MICHELE. 1731. *Commentariolus de laudibus Patavii*, in Muratori, *Rer. Ital. Script.* [56], Vol. XXIV, Milan.

67. Scardeonius, Bernardinus. 1560. *De antiquitate urbis Patavii & claris civibus Patavis . . .* (Basle).

67a. Smith, Thomas. 1696. *Catalogus librorum manuscriptorum Bibliothecae Cottonianae* (Oxford).

68. Squarzafichus. 1711. *Francisci Petrarchae vita ac testamentum* (Rudolstadt).

69. Tanner, Thomas. 1748. *Bibliotheca Britannico-Hibernica: sive, de scriptoribus qui in Anglia, Scotia, et Hibernia ad saeculi XVII initium floruerunt, literarum ordine juxta familiarum nomina dispositis commentarius* (London).

70. Thorndike, Lynn. 1934. *A History of Magic and Experimental Science* (New York, Columbia University Press) Vols. III and IV.

70a. —— 1949. *The Sphere of Sacrobosco and its Commentators* (Chicago, University of Chicago Press).

71. ——, and Pearl Kibre. 1963. *A Catalogue of Incipits of Medieval Scientific Writings in Latin* (Medieval Academy of America, Publication No. 29, rev. ed., London).

72. Tiraboschi, Girolamo. 1823-1825. *Storia della letteratura italiana* (9 v. in 12, Venice, Tip. Molinari).

73. Usher, Abbott Payson. 1954. *A History of Mechanical Inventions* (Cambridge, Mass., Harvard University Press) 2nd edition.

73a. Uzielli, Gustavo. 1894. *Ricerche intorno a Leonardo da Vinci* (Serie prima, Torino, Loescher).

74. Valentinelli, Joseph. 1871. *Biblioteca manuscripta ad S. Marci Venetiarum* (Venice) IV.

75. Vallentin, Antonina. 1938. *Leonardo da Vinci, The Tragic Pursuit of Perfection* (New York, Viking Press).

76. Vedova, Giuseppe. 1832. *Biografia degli scrittori padovani* (Padua, Tip. della Minerva).

76a. Von Bertele, H., and E. Neumann. 1965. *Die Kaisermonument-Uhr. Monographie einer historisch bedeutungsvollen Figurenuhr aus der Spätzeit Kaiser Karls V (1500-1588)* (Lucerne, Buchdruckerei Keller & Co AG).

77. Wickersheimer, Ernest. 1926. *Dictionnaire biographique des médecins en France au moyen âge* (Paris).

78. Wisłocki, Władysław. 1877-1881. *Katalog rekopisów Biblijoteki Uniwersytetu Jagiellońskiego* (Cracow) I.

79. Woodbury, Robert S. 1958. *History of the Gear-Cutting Machine. A Historical Study in Geometry and Machines* (Cambridge, Mass., Technology Press).

80. Zinner, Ernst. 1938. *Leben und Wirken des Johannes Müller von Königsberg, genannt Regiomontanus* (Munich).

PERIODICALS

80a. (Anonymous). 1961. "The Dondi Clock." *Horological Journal* (June): pp. 354-356.

81. Baillie, Granville Hugh. 1934, April and May. "Giovanni de'Dondi and his Planetarium Clock in 1364." *Horological Journal,* pp. 1-12.

82. Bedini, Silvio A. 1956. "An Old Timer." *Hobbies* (May), pp. 36-37.

83. Belgrano, L. T. 1868. "Degli antichi orologi pubblici d'Italia." *Archivio storico italiano,* ser. 3, 7: pp. 35-38.

84. Birkenmajer, L. A. 1893. "Marcin Bylica z Olkusza oraz narzędzia astronomiczne, które zapisał Uniwersytetowi Jagiellónskiemu w r. 1493." *Proceedings of the Polish Academy of Science and Letters, Section of Mathematics and Science,* ser. II, 5.

85. Caffi, Michele. 1876. "Il Castello di Pavia." *Archivio storico Lombardo,* III, 3, (30 Sept.), pp. 543-559.

86. Casati, C. 1871. "Resoconto." *Archivio storico italiano* 3.

87. Curti, Orazio. 1963. "Giovanni Dondi dall'Orologio e il suo astrario." *Museoscienza* 17: pp. 12-15.

88. Dell'Acqua, Carlo. 1865. "Il Palazzo ducale Visconteo in Pavia e Francesco Petrarca." *Archivio storico italiano,* ser. 3, 1: pp. 16-18.

89. (De) Dondi dall'Orologio, Monsignor Francesco Scipione Marchese. 1789. "Memoria di Monsignor . . . Notizie sopra Jacopo e Giovanni Dondi dall'Orologio." *Saggi scientifici e letterari dell'Accademia di Padova* (Padua) 2: 469-494.

90. De Maisieres, Philippe. 1757. "Songe du viel Pelerin", Lebeuf, *Actes de l'Académie des Inscriptions et Belles Lettres* 16.

91. Escosura y Morrogh, Luis de la. 1888. "El Artificio de Juanelo y el puente de Julio Cesar." *Memorias de la Real Academia de ciencias exactas fisicas y naturales de Madrid* 13, 2: pp. 19-29.

92. Falconet, C. 1745. "Dissertation sur les anciennes horloges et sur Jacques de Dondis, surnommé Horologius." *Mémoires de l'Académie des Belles-Lettres* 34.

93. Fonteñla, Luís Montañés. "Les Relojes del Emperador. Replanteo provisional de este tema." Published as a reprint from *Cuadernos de relojería* (Madrid) No. 18, but stated "no publicado."

94. Gloria, Andrea. 1896. "I due orologi meravigliosi inventati da Jacopo e Giovanni Dondi." *Atti del Reale Istituto Veneto di Scienze, Letteri ed Arti,* ser. VII, 7.

95. —— 1896-1897. "L'Orologio inventato da Jacopo Dondi." *Atti del Reale Istituto Veneto di Scienze, Lettere ed Arti,* ser. VII, 8 (reprinted separately Venice, Tipo, Ferrari, 1897, 12 pp.).

96. Lazzarini, Vittorio. 1925. "I libri, gli argenti, le vesti di Giovanni Dondi dall'Orologio." *Bollettino del Museo Civico di Padova,* ser. 1, 1: pp. 11-36.

97. Lloyd, H. Alan. 1955. "Giovanni de Dondi's Horological Masterpiece 1364." *La Suisse Horlogère.* International Edition (July) 2: pp. 49-71.

98. —— 1961. "Il Capolavoro d'orologeria di Giovanni de'-Dondi." *La Clessidra* 17: pp. 9-11.

98a. —— 1964. "The Reproduction of Giovanni Dondi's Astronomical Clock of 1364," *La Suisse Horlogère,* International Edition, 1 (March): pp. 45-58.

99. Maddison, Francis. 1963. "Early Astronomical and Mathematical Instruments. A Brief Summary of Sources and Modern Studies." *History of Science, An Annual Review of Literature, Research and Teaching* (Cambridge) 2: pp. 17-50.

100. Michel, Henri. 1960. "L'Horloge de Sapience et l'histoire de l'horlogerie." *Physis, Rivista di storia della scienza* 2, 4: pp. 291-298.

101. Morpurgo, Enrico. 1962. "Raffronto tra l'astrario e il planetario del Dondi." *La Clessidra* 9 (Anno XVIII Sept.).

101a. —— 1964. "Giovanni de' Dondi, un grande dimenticato." *Clessidra* 22, 6 (June): pp. 12-15.

101b. —— 1965. "Ancora una volta: l'Astrario del Dondi." *Clessidra* 21, 2 (February): pp. 12-13.

101c. Olszewski, Eugeniusz. 1965. "Outline of the Development of Polish Science." *Organon,* 2: pp. 249-259.

102. Parisi, Bruno. 1952. "I Manoscritti di Giovanni Dondi dall' Orologio." *La Clessidra* 2 (Anno VIII, Feb.): pp. 17-18.

103. Pasini. "Codices Mss. Bibliothecae Taurin." *Athenaei* 2, p. 475.

103a. Pedretti, Carlo. 1952. "Nuovi documenti riguadanti Leonardo da Vinci, I., Il Codice di Benvenuto di Lorenzo della Golpaia." *Sapere* (15 April), pp. 57-60.

103*b*. —— 1957. 'Il codice di Benvenuto di Lorenzo della Golpaia." *Travaux d'humanisme et renaissance, Studi Vinciani* 27 : pp. 23–33.

103*c*. PIPPA, LUIGI. 1964. "La ricostruzione dell'Astrario del Dondi." *Clessidra*, **20**, 11 (November) : pp. 22–29.

104. POULLE, EMMANUEL. 1961. "L'Equatoire de Guillaume Gilliszoon de Wissekerke." *Physis* **3**, 3 : pp. 223–251.

105. POULLE, EMMANUEL, and FRANCIS MADDISON. 1963. "Un Equatoire de Franciscus Sarzosius." *Physis* **5**, 1 : pp. 43–64.

106. PRICE, DEREK J. DE SOLLA. 1955–1956. "Clockwork Before the Clock." *Horological Journal*, December 1955 : pp. 810–814; January 1956 : pp. 31–34, 35.

107. —— 1956. "Two Medieval Texts on Astronomical Clocks." *Antiquarian Horology* 10: p. 156.

108. —— 1956. "The Prehistory of the Clock." *Discovery*, April : pp. 153–157.

109. —— 1959. "On the Origin of Clockwork, Perpetual Motion Devices and the Compass" (Contributions from the Museum of History and Technology), *U. S. National Museum Bulletin No. 218* (Washington, D. C., Smithsonian Institution) **6**: pp. 81–112.

110. —— 1962. "Unworldly Mechanics." *Natural History* **71**, 3 (March) : pp. 8–17.

111. —— 1958. "Leonardo da Vinci and the Clock of Giovanni de'Dondi." *Antiquarian Horology* **2**, 7: pp. 127–128. Also Letters, 2, 10: p. 199; 2, 11: p. 222.

112. PRZYPKOWSKI, TADEUSZ. 1961. "Premières cartes modernes du ciel." *Archives internationales d'histoire des sciences* **14**, 56-57 : pp. 305–313.

112*a*. RETI, LADISLAO. 1959. "Non si volta chi a stella e fisso. Le 'imprese' di Leonardo da Vinci." *Bibliothèque d'Humanisme et Renaissance, Travaux et documents* **21**: pp. 7–54.

112*b*. —— 1965. "A Postscript to the Filarete Discussion. On Horizontal Waterwheels and Smelter Blowers in the Writings of Leonardo and Juanelo Turriano." *Technology and Culture* **6**, 3: pp. 428–441.

113. RIVERO, CASTO MARIA DEL. 1936. "Nuevos documentos de Juanelo Turriano." *Revista española de arte* (primer trimestre), pp. 17–21.

113*a*. SÁNCHEZ MAYENDIS, JOSÉ CRISTÓBAL. 1962. "Los Retrados de Juanelo." *Cuadernos de Relojeria* **10** (Jan.–Feb.–Mar.) : pp. 27–29.

114. SIMONI, ANTONIO. 1952. "Giovanni de' Dondi e il suo orologio dei pianeti." *La Clessidra* **8**, 2: pp. 7–16.

115. SOLMI, EDMONDO. 1904. "La Festa del Paradiso di Leonardo da Vinci e Bernardo Bellincione (13 gennaio 1490)." *Archivio storico Lombardo* **1** (Anno XXXI) : pp. 75–89.

116. SOTHEBY, WILKINSON & HODGE. . . . 1919. *Catalogue of the Extensive and Interesting Library, the Property of the Late Sir Frank Crisp, Bart., of Friar Park, Henley-on-Thames . . .* [*sold on*] 17 November . . . [*to*] 20 November . . . 1919.

116*a*. SPENCER, ELEANOR P. 1963. "L'Horloge de Sapience." Bruxelles, Bibliothèque royale, MS. IV. III. *Scriptorum* **17**: pp. 277–299.

117. THORNDIKE, LYNN. 1936. "Milan Manuscripts of Giovanni de' Dondi's Astronomical Clock and of Jacopo de' Dondi's Discussion of Tides." *Archeion, Archivio di storia della scienza* **18**, 4: pp. 308–317.

117*a*. —— 1941. "Invention of the Mechanical Clock about 1271 A.D." *Speculum* **16**: pp. 242–243.

117*b*. WIEDEMANN, EILHARD. 1913. "Ein Instrument das die Bewegung von Sonne und Mond darstellt, nach al Biruni." *Der Islam* **4**: pp. 5–13.

118. ZINNER, ERNST. 1957. "Die Planetenuhren von Dondi und Regiomontan." *Die Uhr* 21 pp.

DOCUMENTS

ARCHIVIO DI STATO, MILAN

119. *Missive* (Facino da Fabriano to Duca), 17 April, 1456, unnumbered. (Document IX.)

120. *Missive*, Reg. 32, folio 17, verso, 9 April, 1456.

121. *Lettere* (Facino da Fabriano to Duca), 1 May, 1456.

122. *Missive Ducale*, Anni 1453–1464, folio 124, transcript, unnumbered separate sheet.

123. *Missive* (Cicco Simonetta to Conte Bolognini de Attendolis), 28 May, 1459, separate sheet.

124. *Registro delle Missive*, N. 47, 6 January, 1460, folio 110 bis.

125. *Missive*, N. 55, folio 315, transcript (Studi-Libri-Astrolabio-Quadrante), 26 September, 1463.

126. *Missive*, Registro N. 68, folio 31, 5 November, 1464 (Francesco Sforza to Conte Bolognino de Attendolis).

127. *Missive*, Registro N. 98, folio 164, 14 February, 1471 (Galeazzo Maria Sforza to Francesco Sforza).

128. *Carteggio Generale*, 6 November, 1494 (Ludovico Maria Sforza to Gualtiero Bescapé).

129. *Feudi, Rosate e Registri Panigarola*, I c 182 t°.

130. *Autografi, Medici* (Ambrogio da Rosate).

131. *Registri Ducale*, N. 121. Autografi C. 98—Bramante, 5 March, 1495 (Jacopo da Pusterla to Ludovico Maria Sforza).

132. *Missive*, Registro Ducale, Libro 193, folio 211 verso, 6 March, 1495 (Duca to Custodio del Castello).

133. *Missive*, Registro Ducale, Libro 37, folio 90.

ARCHIVIO DI STATO DI MODENA

134. *Carteggio degli Ambasciatori Estensi in Milano*, Busta 11ª (Lettera da Blanchino da Palude to Duca) 4 March, 1495.

BIBLIOTECA SEMINARIO VESCOVILE DI PADOVA

135. *Lettera autografa di Francesco Petrarca al Giovanni de' Dondi*, 1380.

136. *Codice N. 358*. Saec. xiv-xv, Reply from de' Dondi to Petrarca; Senilium liber XIII, cc. 15 and 16, two letters from Petrarca to de' Dondi.

ARCHIVIO DI STATO, PAVIA

137. *Registro delle bollette e de'salari mensuali del comune, compilato 1399* (Comune di Pavia), folio 56.

MANUSCRIPT BOOKS

138. BOSSI. 1381. *Storia pavese* (Milan, Archivio di Stato).

139. DAL POZZO, SIMON. 1548. *Libro grosso dell 'estimo generale* (Milan, Archivio di Stato).

140. DA VINCI, LEONARDO. *Codice Atlantico* (Milan, Biblioteca Ambrosiana).

141. —— *Manuscripts B, L and M* (Paris, Institut de France).

142. —— *Arundel Manuscript* (London, British Museum), No. 263, Sched. No. 7416.

143. DECEMBRIUS, UBERTUS. *De Republica* (Milan, Biblioteca Ambrosiana).

144. (DE' DONDI, ——). *Familia nostra in Civitate Cremonensi florens ex illa factione populari, ut assolet, pulsa fuit anno 1251, ac se Patavium contulit* (Padua, Biblioteca del Museo Civico), Archivio Privato Famigliare Dondi dall'Orologio.

145. (TROTTO, GIACOMO). *Raccolta di vari monumenti istorici e varie narrazioni. Feste in Milano nel 1490* (Paradiso). (Modena, Biblioteca Estense) Cod. Ital. N 421 a J. 421.

INDEX

67